Molecular Medicine

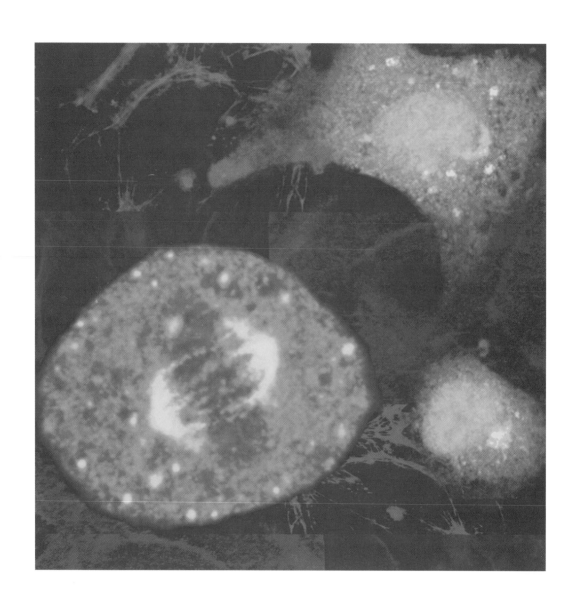

LECTURE NOTES ON

Molecular Medicine

JOHN BRADLEY

BMedSci, DM, MRCP
University of Cambridge School of Clinical Medicine
Addenbrooke's Hospital, Cambridge, UK

DAVID JOHNSON PhD

Boyer Center for Molecular Medicine
Yale University School of Medicine
New Haven, Connecticut, USA

DAVID RUBENSTEIN

MA, MD, FRCP
University of Cambridge School of Clinical Medicine
Addenbrooke's Hospital, Cambridge, UK

b

**Blackwell
Science**

© 1995 by
Blackwell Science Ltd
Editorial Offices:
Osney Mead, Oxford OX2 0EL
25 John Street, London WCIN 2BL
23 Ainslie Place, Edinburgh EH3 6AJ
238 Main Street, Cambridge
  Massachusetts 02142, USA
54 University Street, Carlton
  Victoria 3053, Australia

Other Editorial Offices:
Arnette Blackwell SA
  1, rue de Lille, 75007 Paris
  France

Blackwell Wissenschafts-Verlag GmbH
  Kurfürstendamm 57
  10707 Berlin, Germany

  Feldgasse 13, A-1238 Wien
  Austria

First published 1995

Set by Excel Typesetters, Hong Kong
Printed and bound in Great Britain
at the Alden Press Limited,
Oxford and Northampton

DISTRIBUTORS

Marston Book Services Ltd
PO Box 87
Oxford OX2 0DT
(*Orders*: Tel: 01865 791155
          Fax: 01865 791927
          Telex: 837515)

North America
Blackwell Science, Inc.
238 Main Street
Cambridge, MA 02142
(*Orders*: Tel: 800 215-1000
          617 876-7000
          Fax: 617 492-5263)

Australia
Blackwell Science Pty Ltd
54 University Street
Carlton, Victoria 3053
(*Orders*: Tel: 03 9347-0300
          Fax: 03 9349-3016)

A catalogue record for this title
is available from the British Library

ISBN 0–632–03851–9 (BSL)
ISBN 0–632–03853–5
(International Edition)

Library of Congress
Cataloging-in-Publication Data

Bradley, John.
  Molecular medicine/John Bradley,
  David Johnson, David Rubenstein.
      p.   cm. — (Lecture
  notes)
  Includes index.
  ISBN 0–632–03851–9
  1. Molecular genetics—Outlines,
syllabi, etc.   2. Molecular biology—
Outlines, syllabi, etc.   3. Medical
genetics—Outlines, syllabi, etc.
I. Johnson, David, Ph.D.
II. Rubenstein, David.   III. Title.
IV. Series.   V. Series: Lecture notes
series   (Blackwell Science)
      [DNLM: 1. Genetic Techniques—
laboratory manuals.   2. Genetics,
Biochemical—laboratory manuals.
QU 25 B811m   1995]
QH442.B73   1995
616'.042—dc20
DNLM/DLC
for Library of Congress   95–8412
                                      CIP

Contents

Preface

Molecular medicine is a novel, exciting and important subject which is made difficult by an unfamiliar language. In the first part of this book we have attempted to introduce the discipline to the new reader, using an extensively illustrated text to explain the common terms and their applications. The second part of the book outlines how molecular biology is used to define the molecular basis of disease. Finally, we describe how this is beginning to translate into benefits for patients.

We would like to thank Kathleen Gillespie, Richard Sandford and Phil Roberts for reading the text. We are particularly grateful to Jordan Pober for his comments and advice. Brad Amos generously gave his expert help in preparing the confocal micrographs for the cover and chapter headings.

Abbreviations

abl	Abelson murine leukaemia virus
ADA	adenosine deaminase
AIDS	acquired immunodeficiency syndrome
ATP	adenosine triphosphate (the prefix 'd' indicates deoxy, 'dd' indicates dideoxy)
β-gal	β-galactosidase
bcr	break point cluster region
bp	base pair
cAMP	cyclic adenosine monophosphate
CD	cluster of differentiation
cDNA	complementary DNA
CFTR	cystic fibrosis transmembrane conductance regulator
C_H	constant domain of the immunoglobulin heavy chain
C_L	constant domain of the immunoglobulin light chain
cM	centimorgan
contigs	contiguous sequences
CsCl	caesium chloride
DEAE	diethylaminoethyl
DEPC	diethylpyrocarbonate
DHFR	dihydrofolate reductase
DMS	dimethylsulfoxide
DNA	deoxyribonucleic acid
DNase	DNA degrading enzyme
EDTA	ethylenediamine tetra-acetic acid
eIF	elongation factor
EMSA	electromobility shift assay
ES	embryonic stem
EtBr	ethidium bromide
FISH	fluorescence *in situ* hybridization
G	Giemsa
G-CSF	granulocyte colony-stimulating factor
G1	Gap 1 in eukaryotic cell cycle
G2	Gap 2 in eukaryotic cell cycle
G3	Gap 3 in eukaryotic cell cycle

GM-CSF	granulocyte-macrophage colony-stimulating factor
H	histone
HGPRT	hypoxanthine guanine phosphoribosyl transferase
HIV	human immunodeficiency virus
HLA	human leukocyte antigen
HTF	*Hpa*II tiny fragments
HSV-1	herpes simplex virus type-1
IDDM	insulin dependent diabetes mellitus
IEF	isoelectric focusing
Ig	immunoglobulin
IPTG	isopropyl β-D-thiogalactopyranoside
IVS	intervening sequence
kbp	thousand base pairs
lacZ	gene encoding a subunit of the enzyme β-galactosidase
LDL	low density lipoprotein
lod	logarithm of the odds
luc	luciferase
Mbp	million base pairs
MCS	multiple cloning site
MHC	major histocompatibility complex
mAb	monoclonal antibody
mRNA	messenger ribonucleic acid
OD	optical density
OKT3	orthoclone anti-CD3 monoclonal antibody
ori	origin of DNA replication
p	inserted between two nucleotides (e.g. ApT) indicates a phosphodiester bond
p	used to denote chromosomal location indicates the short (petit) arm
PCR	polymerase chain reaction
PFGE	pulsed field gel electrophoresis
PKD	polycystic kidney disease
PNA	peptide nucleic acids

Pol	polymerase
q	used to denote chromosomal location indicates the long arm
R-M	restriction modification
RFLP	restriction fragment length polymorphism
RNA	ribonucleic acid
RNase	RNA degrading enzyme
rRNA	ribosomal ribonucleic acid
RT	reverse transcriptase
S	Svedberg unit
SV40	Simian virus 40
t-PA	tissue-type plasminogen activator
Taq DNA Pol	Thermus aquaticus DNA polymerase
tRNA	transfer ribonucleic acid
TSC	tuberous sclerosis
UTR	untranslated region
UV	ultraviolet
V_H	variable domain of the immunoglobulin heavy chain
V_L	variable domain of the immunoglobulin light chain
VNTR	variable number tandem repeats
X-gal	5-bromo-4-chloro-3-indolyl-β-D-galactopyranoside
YAC	yeast artificial chromosome

Source of restriction enzymes

*Afl*II	*Anabaena flos-aquae*
*Alu*I	*Arthobacter luteus*
*Bam*HI	*Bacillus amyloliquefaciens* H
*Eco*RI	*Escherichia coli*
*Hinc*II	*Haemophilus influenzae* Rc
*Hind*III	*Haemophilus influenzae* Rd
*Hpa*II	*Haemophilus parainfluenzae*
*Nar*I	*Nocardia argentinensis*
*Not*I	*Nocardia otitidis-caviarum*
*Pst*I	*Providencia stuartii*
*Sma*I	*Streptomyces caespitosus*
*Taq*I	*Thermus aquaticus*
*Xba*I	*Xanthomonas badrii*

Basic principles

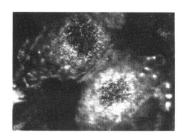

Prokaryotic organism–simple, single-cell life form that lacks a distinct nucleus. Examples include bacteria and certain algae.

Molecular biology is a fascinating subject.

The basic techniques of making, breaking and joining deoxyribonucleic acid (DNA), which have been adopted by molecular biologists, are all used by viruses and prokaryotic organisms as a means of ensuring their survival.

It is the definition of the structure and function of DNA, together with an understanding of these methods of manipulating DNA, that has led to the remarkable breakthroughs in molecular biology in recent decades.

In this chapter we will describe in detail how information is stored in DNA, and how this information is used by the cell.

Organisms are made of cells

Viruses are small cellular parasites that generally consist of DNA or ribonucleic acid (RNA) with a protein coat. In some complex viruses a membrane surrounds the protein coat.

Living organisms are composed of cells. Some organisms, including bacteria, algae and yeasts, exist as single cells, whereas plants and animals consist of collections of cells. New cells, required for growth of an existing organism or the formation of new organisms, arise by division of existing cells.

1

Cell functions depend on proteins

All cellular functions depend on proteins, which consist of chains of amino acids. Only 20 different amino acids are commonly found in the proteins of all organisms.

The links in a chain of amino acids are termed peptide bonds, and the chains themselves are called polypeptides. Proteins contain one or more polypeptides, and the structure and function of each protein depends on the sequence of amino acids making up the polypeptide chains.

Proteins have many diverse functions. They maintain cell structure and provide motility, act as intra- and extracellular messengers, and bind and transport molecules, including oxygen, lipids and other proteins. Many proteins are enzymes which catalyse (accelerate) chemical reactions. Almost all chemical reactions, including those involved in the synthesis of fats and carbohydrates, are catalysed by enzymes.

Some proteins, e.g. the enzymes involved in glucose metabolism, occur in most cells. In contrast, cells in multicellular organisms may become specialized and produce certain proteins that provide them with highly specific functions. Cells that produce particular proteins are often grouped together to form complex tissues or organs. For example, muscle cells produce proteins, including tropomyosin and myosin, which are involved in the formation of muscle filaments, islet cells of the pancreas synthesize the polypeptide hormone insulin, and liver cells contain enzymes found exclusively in the liver, such as those required for the conjugation of bilirubin into water-soluble forms.

DNA contains the information needed to encode proteins

Cells therefore need:
• the information to produce proteins in a regulated fashion;
• the ability to convey this information to daughter cells during cell division.

The key to these requirements is provided by the *DNA double helix*, which contains two strands of DNA held together by weak chemical interactions.

The strands complement each other—the sequences of bases on one strand can be determined from the sequence of the other strand. During cell division each strand independently forms a new complementary strand, and the DNA helix is able to direct its own duplication.

BASIC DNA STRUCTURE

Each strand of DNA has a backbone of sugars and phosphates, with a nitrogen containing base attached to each sugar. Four different bases are found in DNA. Cytosine (C) and thymine (T) are pyrimidines which contain one nitrogenous ring, whereas adenine (A) and guanine (G) are purines which contain two. The bases from each strand are linked together to form the 'rungs' inside the helix in such a way that A can only pair with T, and C can only pair with G.

The sequence of bases in a DNA molecule carries the information that specifies the order of amino acids along a polypeptide chain. Each of the 20 amino acids is encoded by coding units, or codons, which consist of three consecutive bases. Reading this code, and translating it into protein, requires ribonucleic acid (RNA).

In RNA the sugar is ribose, uracil replaces T and the resulting nucleic acid is single stranded.

A segment of DNA that carries the information needed to encode a specific polypeptide is known as a gene. To retrieve this information a single-stranded messenger RNA (mRNA) copy of the gene is made, and the sequence of bases in the mRNA is then translated into a linear sequence of amino acids, composing a polypeptide. Genetic information is therefore stored in cells in DNA. During the expression of a gene, a segment of DNA is first transcribed into RNA, and then translated from RNA into protein. During cell division DNA replicates itself to form two identical DNA helices.

DNA in eukaryotic organisms is organized into chromosomes within the cell nucleus

Living things may be divided into prokaryotes and eukaryotes. Prokaryotic organisms are simple, single cell life forms that lack a distinct nucleus. Examples include bacteria and certain algae. The cells in eukaryotic organisms contain nuclei. Eukaryotes may be single cell life forms such as yeasts, or complex multicellular organisms such as plants and animals. DNA within the nucleus of eukaryotes is organized into chromosomes. Each chromosome contains an extensively folded, DNA double helix.

DNA

DNA is composed of three principal structures:
• bases;

- sugars;
- phosphates.
  These are linked together by three principal types of bonds:
- covalent;
- hydrogen;
- ester.

The players

Bases

The bases in DNA are nitrogen containing rings (the nitrogen makes these molecules basic). Pyrimidines (C, T) have one ring, whilst purines (A, G) have two.

Base – a molecule that can combine with hydrogen ions in solution.

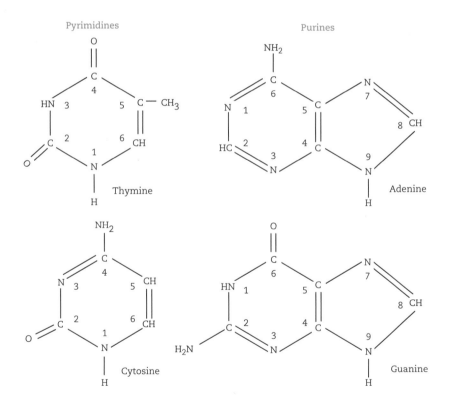

Pyrimidines   Purines

Thymine   Adenine

Cytosine   Guanine

Sugars

The sugars in DNA are pentoses (sugar molecules containing five carbon atoms). In DNA the pentose is always deoxyribose, indicating that it lacks an oxygen molecule that is present in ribose, the parent compound.

Ribose could not fit into a DNA helix as there is insufficient room for the 2'-OH group.

$$OH-\overset{5'}{\underset{|}{C}}-H_2$$

Ribose

$$OH-\overset{5'}{\underset{|}{C}}-H_2$$

Deoxyribose

By convention the carbon atoms are labelled by primed numbers (1' to 5') when part of a nucleotide. This labelling is important in understanding how the DNA molecule is assembled.

Phosphates

Acid – a molecule that releases a hydrogen ion in solution.

The phosphates in DNA are either mono-, di- or tri-phosphates. The acidic character of nucleic acid is due to the presence of phosphate esters, which are relatively strong acids.

At neutral pH they dissociate from hydrogen ions, and are thus normally referred to in their ionized form:

Monophosphate   Diphosphate   Triphosphate

The bonds

Covalent bonds

A covalent bond exists between atoms that share electrons in their outermost shell. The bonding electrons move freely around both nuclei which are held close together in a strong bond – energy is released when the bonds are formed, and the same amount of energy is required to break the bond.

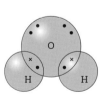

In water each hydrogen atom shares its electron (x) with an electron from the outer shell of the oxygen atom (•). The hydrogen atom thus fills its outer (first) shell with two electrons, and the oxygen atom fills its outer (second) shell with eight

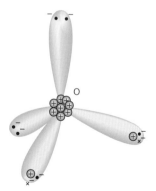

Outer orbitals have characteristic shapes. The four outer orbitals of an oxygen atom in water point outwards, forming the corners of a tetrahedron

Hydrogen bond

A hydrogen atom can usually only form one covalent bond with another atom. A covalently bonded (*electron depleted*) hydrogen atom can, however, form a weak electrostatic inter-action (*hydrogen bond*) with an electronegative (*electron rich*) atom (usually nitrogen or oxygen), e.g.

N ▬▬ H ---- O

Covalent Hydrogen
bond   bond

Ester bond

An ester bond involves covalent bonding, and is formed when an alcohol and an acid unite with elimination of water.

BOND STRENGTH

The strength of the bonds is important in understanding the stability of different parts of the final DNA molecule. Strong covalent bonds link nucleic acids in a single DNA strand, whereas weaker hydrogen bonds hold two DNA strands together.

The formation of DNA

Base + Sugar = nucleoSide

The 1′ carbon of pentose ring is attached to nitrogen 1 of pyrimidine or nitrogen 9 of purine

nitrogen 1 of pyrimidine   or nitrogen 9 of purine.

Base + sugar + phosphaTe = nucleoTide
Phosphate is attached to the 5′-carbon of the pentose ring

NUCLEOTIDES AS ENERGY STORES

Nucleotides may have either one, two or three phosphates attached. In addition to forming the building blocks of DNA, the nucleotide di- and triphosphates are important stores of chemical energy; cleavage of the terminal phosphate bond releases energy which is used to drive cell functions. Adenosine triphosphate is the most widely used energy carrier in the cell.

Nucleotides join together to form *nucleic acid*
The hydroxyl group attached to the 3′-pentose carbon of one nucleotide forms an ester bond with the phosphate of another molecule, eliminating a water molecule. The link between nucleotides is known as a phosphodiester bond. Thus, one end of a DNA strand has a sugar residue in which the 5′-carbon is not linked to another sugar residue (the 5′ end), whereas at

Phosphodiester linkage

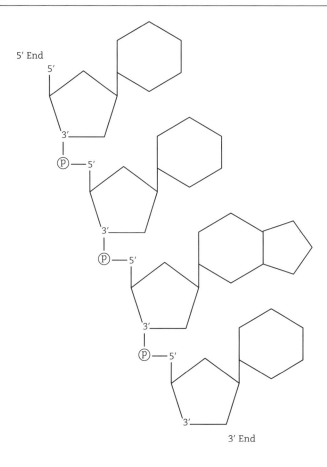

This simple terminology is fundamental to understanding descriptions of how DNA replicates and is expressed.

the other end the 3'-carbon lacks a phosphodiester bond (the 3' end).

DNA structure

The DNA helix

In the 1950s X-ray diffraction data suggested that DNA is helical (Fig. 1.1a). In addition, biochemical data showed that the amount of A in DNA always equalled that of T, whilst the amount of G equalled that of C. These observations led Watson and Crick to propose the double-helical structure of DNA, which could account for the physical properties of DNA and its replication in the cell.

The 'backbone' on the outside of the helix consists of alternating sugars and phosphates. The bases are attached to the sugars and form the 'rungs' of the helix.

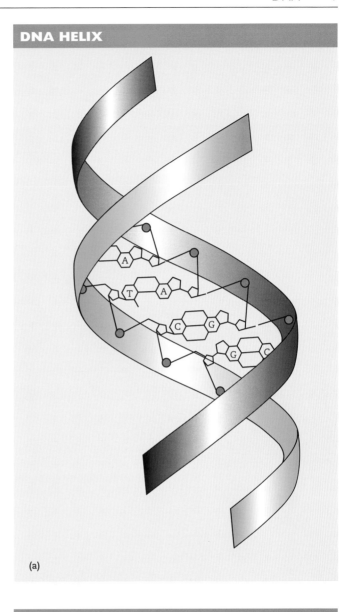

DNA HELIX

(a)

AT/GC BASE PAIRING

(b)

Fig. 1.1 (a) Diagrammatic representation of the DNA helix. (b) AT/GC
base pairing.

MAJOR/MINOR GROOVES

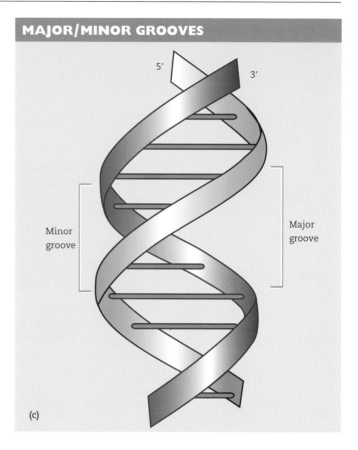

Minor groove

Major groove

(c)

Fig. 1.1 (*Continued*) (c) The major and minor grooves of the DNA helix.

CLEAVAGE OF DNA BONDS

The relative weakness of the hydrogen bonds holding the base pairs together is demonstrated by 'melting' the DNA. At increased temperatures the two strands separate: the DNA melts.

The bonds holding the backbone of the helix together are stronger and do not melt, but can be cleaved by enzymes derived from bacteria which cut the backbone at specific sites.
  Bacteria use these enzymes as protective devices to degrade foreign DNA. They restrict the growth of viruses which infect bacteria (bacteriophages), and are known as restriction enzymes.

Bacteriophage is a virus that infects bacteria, sometimes referred to as a phage.

As the distance between the sugar–phosphate backbone is fixed by the diameter of the helix, only two types of base pairs (AT or CG) can fit, explaining the constant regularity in the ratios between base pairs (A = T and G = C) (Fig. 1.1b).

The strands are antiparallel (their 5',3'-phosphodiester links run in opposite directions) and complementary (because of base pairing the chains complement each other). The sequences of bases on one strand can thus be deduced from the sequences of bases on the other, and each strand independently carries the information needed to form a double helix.

DESCRIBING A DNA SEQUENCE

It is conventional to describe a DNA sequence by writing the sequence of bases in one strand only, and in the 5' → 3' direction. When identifying just two neighbouring bases in a sequence it is usual to insert 'p' between them to denote an intervening phosphodiester bond (e.g. ApT). This is distinct from AT which indicates a hydrogen-bonded base pair on complementary strands.

The DNA helix can take on several conformations. The most common form is *B-DNA*, in which the helix is right handed and has just over 10 bp per helical turn. There are two unequal grooves, the major and minor grooves (Fig. 1.1c).

A-DNA is a right-handed helix which is shorter and wider than B-DNA. The phosphate groups bind fewer water molecules, and its formation is thus favoured by dehydration.

Z-DNA is a left-handed helix in which alternating purines and pyrimidines give rise to a Zigzag appearance to the helix.

Chromatin

The total length of all the strands of DNA in a human cell is ~2 m, all of which needs to be packed into a nucleus a few micrometres in diameter. This is achieved by the formation of a nucleoprotein complex called *chromatin*: acidic phosphates in the backbone of DNA enable it to form ionic bonds with basic lysine and arginine rich proteins known as histones. Coiling of DNA around histone proteins allows long strands to be tightly packed into chromatin.

DNA is first packaged into a *nucleosome* (Fig. 1.2), which consists of eight histone proteins around which a strand of DNA containing 146 bp is wound one and three-quarter times.

The histone protein H1 binds to DNA just next to each nucleosome, and is involved in coiling DNA into chromatin fibres of 30 nm diameter (Fig. 1.3).

There are five major histone proteins termed H1, H2A, H2B, H3 and H4. The core of the nucleosome contains two copies each of H2A, H2B, H3 and H4.

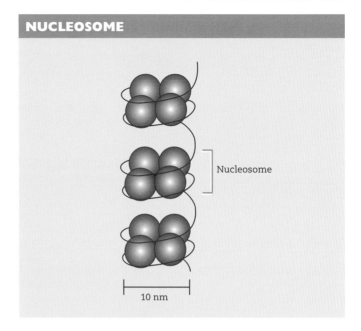

Fig. 1.2 A nucleosome.

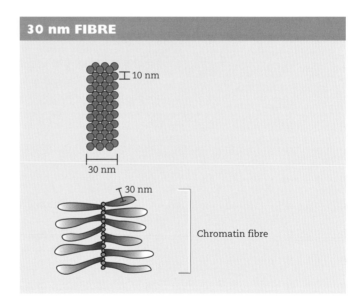

Fig. 1.3 A chromatin 30 nm fibre.

Chromosomes

During cell division chromatin becomes more condensed, and can be recognized in the form of *chromosomes* by light microscopy. During metaphase each chromosome consists of two symmetrical *chromatids*, each containing DNA in which the chromatin fibres are folded in loops around a central scaffold

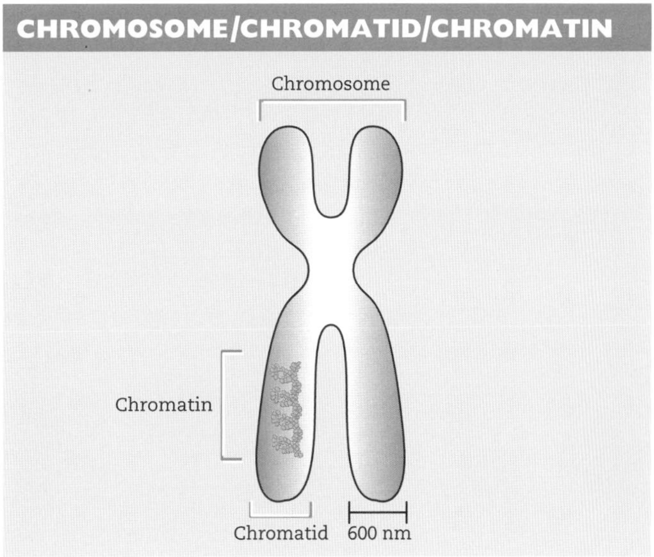

CHROMOSOME/CHROMATID/CHROMATIN

Fig. 1.4 Chromosomes, chromatids and chromatin.

of non-histone acidic protein. The chromatids are attached to each other at the centromere (Fig. 1.4).

Condensed metaphase chromosomes can be subdivided by various treatments which cause the appearance of light and dark bands. For example, staining with Giemsa gives rise to alternating dark- and pale-staining *G bands* (Fig. 1.5). Such banding allows classification of sites on the chromosome according to their location on the short arm (p for petit), or long arm (q), and their position relative to the centromere. For example, the gene that encodes the β-globin chain of haemoglobin (which is abnormal in β-thalassaemia), has been localized to the short arm of chromosome 11, in region 1, band 5, subband 5: written as 11p15.5 (Fig. 1.5).

DNA IN PROKARYOTES

In prokaryotes all the DNA exists in a single molecule which is circular. There are no 5′ or 3′ ends and no histones, and there is no nucleus. The DNA can, however, be induced to supercoil into a compact structure around DNA binding proteins by the enzyme DNA gyrase.

Karyotype

Every species has a specific number and form of chromosomes, which is referred to as a *karyotype*. Human cells contain 46 chromosomes, of which two are sex chromosomes (two X

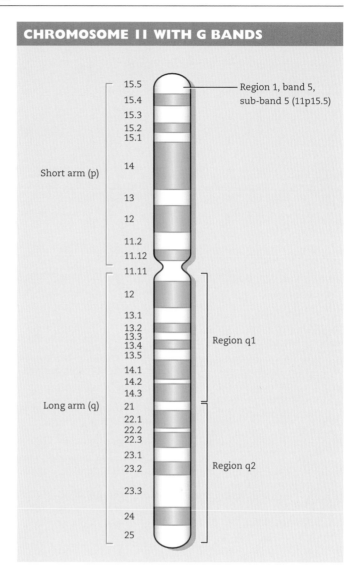

CHROMOSOME 11 WITH G BANDS

15.5

15.4

15.3

15.2
15.1

Region 1, band 5, sub-band 5 (11p15.5)

Short arm (p)

14

13

12

11.2

11.12

11.11

12

13.1
13.2
13.3
13.4
13.5

Region q1

14.1
14.2
14.3

Long arm (q)   21

22.1
22.2
22.3

23.1
23.2

Region q2

23.3

24

25

Fig. 1.5 Chromosome 11 with G bands.

chromosomes in females, an X and a Y chromosome in males), and 44 are autosomes (22 matching pairs numbered 1 to 22) (Fig. 1.6).

Genome

The complete genetic make-up of an individual is referred to as their *genome*. Thus, in human cells the genome is composed of 23 pairs of chromosomes within the nucleus, each of which contains a single, linear, double-helical DNA. The human genome contains approximately 3×10^9 bp (base pairs), and is thought to contain 50 000 to 100 000 different genes, most of

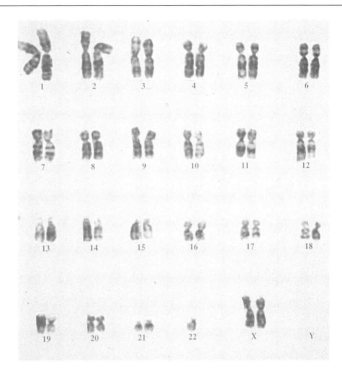

Fig. 1.6 A normal human female karyotype stained with Giemsa. Courtesy of Genetics Laboratories, Addenbrooke's Hospital, Cambridge.

Although mitochondria possess their own genome, the majority of mitochondrial proteins are encoded by nuclear genes.

which encode polypeptides. A small minority of genes encode RNA molecules. In addition to the nuclear genome, eukaryotic cells also contain a small mitochondrial genome which is inherited from the mother. This is because, unlike sperm, eggs have a considerable amount of cytoplasm which contains mitochondria. In humans the mitochondrial genome consists of a 16569 bp circular DNA molecule which encodes proteins essential for mitochondrial structure and function, including oxidative enzymes, together with RNA molecules involved in mitochondrial protein synthesis.

RNA STRUCTURE

DNA carries the information that encodes polypeptides. Reading this code, and translating it into specific proteins, involves RNA.

RNA differs from DNA in the following respects:
- RNA is single stranded (usually);
- the sugar in RNA is ribose rather than deoxyribose;
- RNA has uracil (U) rather than T as one of its pyrimidines.
  RNA exists in three forms.
- Messenger RNA (*mRNA*) is a copy of DNA which encodes a specific amino acid sequence.

The code carried by mRNA is read and translated into protein in ribosomes. Ribosomes are spherical structures in the cytoplasm of cells, composed of ribosomal RNA (rRNA) and ribosomal proteins.

• *Transfer RNA (tRNA)* carries amino acids to ribosomes.

• *Ribosomal RNA (rRNA)* facilitates the interaction between mRNA and tRNA, resulting in the translation of mRNA into protein.

Whilst mRNA is formed as a copy of a gene encoding a specific polypeptide, tRNA and rRNA are formed as the products of genes which actually encode RNA molecules. Multiple copies of genes encoding different tRNA and rRNA molecules occur.

tRNA AND rRNA

tRNA molecules differ according to the amino acid they carry. About 1600 copies of genes encoding different tRNA molecules are dispersed throughout the human genome.

rRNA is formed of different sized subunits, designated 28S, 5.8S, 18S and 5S. The 28S, 5.8S and 18S subunits are formed by processing a large precursor RNA molecule encoded by about 300 copies of a gene which occurs on chromosomes 13, 14, 15, 21 and 22. Copies of the gene encoding the 5S subunit are clustered together on chromosome 1.

Subunits of RNA are commonly designated S, or Svedberg units, which are related to the size and shape of the RNA. Svedberg units are actually measures of the sedimentation rate of molecules centrifuged through a sucrose gradient. This was a common method of analysing macromolecules before gel electrophorsesis (see p. 44) became routine.

DNA REPLICATION AND TRANSCRIPTION: IT'S ALL IN THE GENES

The double-helical structure of DNA provides a mechanism by which nucleic acids can accurately replicate and provide the information for building proteins.

During replication the two chains dissociate and each one serves as a template for the synthesis of two complementary strands of DNA. During gene expression information is retrieved from only one of the two available strands. The segment of DNA containing a gene is first transcribed into a single-stranded mRNA copy which has the same sequence of bases as the sense strand of DNA, and is complementary to the antisense strand. The sequence of bases is then translated into a sequence of amino acids composing a polypeptide (see p. 30).

The sense strand of DNA has the same sequence of bases as the transcribed mRNA (in which U replaces T). The antisense strand carries the complementary sequence of bases.

The flow of genetic information in a cell can thus be summarized as follows.

The direction of synthesis of DNA during replication, or RNA during transcription, is $5' \rightarrow 3'$.

DNA replication

When viruses are taken into consideration, genetic information can also pass from RNA to RNA during replication of some RNA viruses, and from RNA to DNA in retroviruses, which have RNA genomes from which a DNA copy is made during their infectious cycle. The retrovirus genome contains the gene for the enzyme reverse transcriptase which catalyses RNA-dependent synthesis of DNA.

VIRUSES

Viruses are small cellular parasites which generally consist of DNA or RNA within a protein coat or capsid. In some complex viruses a membrane surrounds the protein coat. A virus cannot replicate by itself, but uses the machinery of infected cells to do so.

DNA replication is *semi-conservative* – one strand (half of the original DNA) is retained ('conserved') in the new DNA molecule.

DNA replication

For DNA replication to occur the following are required.
• A DNA *template* containing a region of single-stranded DNA from which a complementary copy is made. (The double-stranded helix must unwind, and each strand then acts as a template.)

The primer is a short strand of RNA which is synthesized on the template at the start of replication, and removed at the end – RNA *primes the synthesis of DNA*.

• A *primer* chain with a free 3'-OH group at the site where replication originates.
• A supply of *triphosphate nucleotides* to attach to the growing chain.
• *DNA polymerases* (Pols) which catalyse the addition of nucleotides to the pre-existing strand of DNA or primer RNA.

DNA Pols produce a link between the inner phophorus of the nucleotide and the 3'-OH group of the primer – elongation occurs in the $5' \rightarrow 3'$ direction.

• The whole process requires *energy* which is supplied by the hydrolysis of nucleotide triphosphates releasing pyrophosphate. The enzyme pyrophosphatase splits the high-energy phosphoanhydride bond in pyrophosphate, which is converted to inorganic phosphate.

DNA REPLICATION

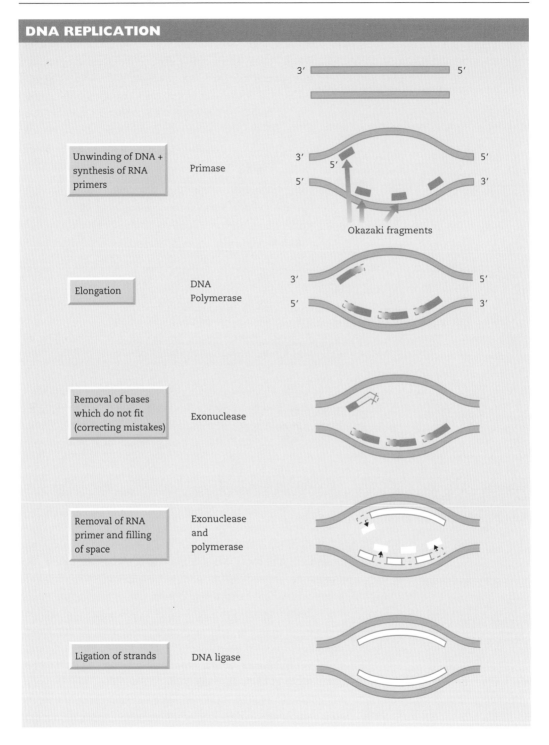

Unwinding of DNA + synthesis of RNA primers	Primase	
Elongation	DNA Polymerase	
Removal of bases which do not fit (correcting mistakes)	Exonuclease	
Removal of RNA primer and filling of space	Exonuclease and polymerase	
Ligation of strands	DNA ligase	

Okazaki fragments

Fig. 1.7 Replication of DNA: driving both ways down a one-way street.

Much of our understanding of how DNA replicates (Fig. 1.7) has come from the study of the bacterium *Escherichia coli*, in which DNA is present as a single circular molecule. Replication starts at a specific site (the site of origin, *oriC*), and proceeds sequentially in opposite directions, even though synthesis can only occur in the 5′ → 3′ direction. This apparent paradox was resolved by the demonstration that synthesis of one strand occurs continuously, whereas the other strand is synthesized in short 5′–3′ fragments (Okazaki fragments), which are then joined together by DNA ligase.

E. coli actually contains three different DNA Pols (I, II and III) which catalyse both the formation and hydrolysis of DNA. The first DNA Pol to be described, DNA Pol I, was found to have three different enzymatic activities:

The 3′–5′ exonuclease is thought to edit each newly attached base, and remove any which are mismatched and will not fit into a double helix.

• a polymerase that catalyses the formation of new phosphodiester bonds in the growing DNA chain;
• a 3′–5′ exonuclease that catalyses the removal of nucleotides from the 3′ end of DNA chains;
• a 5′–3′ exonuclease that cleaves bonds within one chain of a double helix.

5′-3′ exonucleases remove the RNA primer and are thought to repair double-stranded DNA that is damaged.

Pol I can itself be cleaved to yield two fragments, a small fragment containing the 3′–5′ exonuclease and a large fragment (the *Klenow* fragment), with DNA Pol and 5′–3′ exonuclease activity.

The Klenow fragment is used by molecular biologists to accurately synthesize complementary DNA strands.

Klenow fragment

N — 3′–5′ Exonuclease   5′–3′ Exonuclease   Polymerase — C

Most of the DNA in E. coli is in fact synthesized by Pol III, whereas repair of the DNA and removal of the RNA primer is predominantly performed by Pol I.

A similar mechanism for DNA replication occurs in eukaryotic cells, although the process differs in the following respects.
• Polymerases in eukaryotic cells are termed α, β, δ (located in the nucleus) and γ (located in mitochondria), and are less well characterized.
• Because of the size of eukaryotic genomes, replication originates at many sites within each chromosome.
• DNA replication is a virtually continuous process in prokaryotic cells — as soon as the chromosome has duplicated,

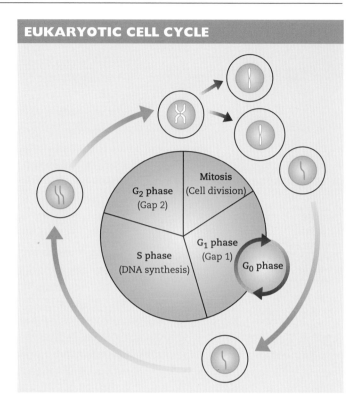

Fig. 1.8 Eukaryotic cell cycle.

the two daughter chromosomes segregate and the cell divides. In eukaryotic cells DNA synthesis and cell division occur at different times.

The whole process of cell growth and division in eukaryotic cells can be divided into different phases, which together make up the cell cycle.

Eukaryotic cell cycle

The eukaryotic cell cycle (Fig. 1.8) consists of two periods: (i) the M (mitotic) period (Fig. 1.9) during which cell division occurs; and (ii) interphase during which cell growth and DNA replication occur. Interphase is further divided into:

- G_1 (Gap 1);
- S (Synthetic);
- G_2 (Gap 2) phases.

DNA synthesis occurs only during the S phase of the cell cycle, which is followed by a gap period (G_2) before cell division (mitosis) occurs. Mitosis is followed by a further gap period (G_1) during which the cell prepares for DNA synthesis. Cells which are not preparing for DNA synthesis and cell

MITOSIS

Interphase Replication of DNA occurs, but chromosomes are not distinguishable by light microscopy. DNA involved in synthesis of RNA is condensed into the nucleolus. The two centrioles duplicate to form daughters

Early prophase The chromosomes appear as long threads. The nucleolus disperses and the centrioles start separating

Middle prophase Chromosomes condense to form chromatids, each containing one of the two DNA molecules that were produced during interphase. The centrioles, which are constructed of microtubules, start to form a spindle as they move to opposite poles of the cell

Late prophase The centrioles reach the poles and are linked by spindle fibres that extend to the centre (equator) of the cell, or attach to the kinetochore (close to the centromere) of each chromatid. The nuclear membrane disperses and disappears

Metaphase The chromatids become aligned on the equator

Early anaphase The two chromatids separate

Late anaphase Each set of chromatids (now a new set of chromosomes) moves to each pole. Separation of the cell cytoplasm (cytokinesis) begins

Telophase Nuclear membranes form around the separated chromosomes, which uncoil and become less distinct. Cytokinesis continues and the spindle disappears

Interphase Following completion of cell division, DNA replication starts again

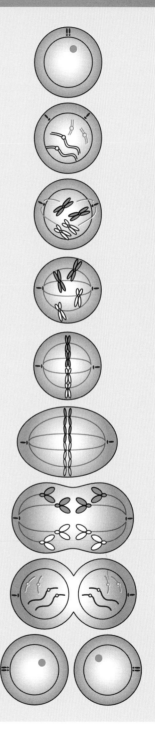

Fig. 1.9 Mitosis.

division may leave G_1 and enter a stage called G_0. Such cells may be metabolically active, but they do not proceed through the cell cycle. G_0 may represent a temporary quiescent state, from which the cell returns to G_1, or a terminally differentiated state. Non-replicating cells such as nerve cells are generally stopped in G_0.

GENE EXPRESSION: MAKING PROTEINS

A *gene* is a sequence of DNA which codes for one polypeptide.

Gene expression involves the transcription of a segment of DNA into RNA, and the translation of RNA into a polypeptide.

Genes carry the code for the amino acids in polypeptide chains. The genetic code is the same in all organisms.

GENETIC CODING

Each of the 20 amino acids (for abbreviations see next section) is encoded by a triplet of bases (*codon*). Since their are four bases there are: $4 \times 4 \times 4 = 64$ possible codons, and single amino acids may be encoded by more than one codon (there is 'redundancy' in the code).

Certain codons specify the beginning and end of polypeptide chains. Methionine, encoded by AUG (or more rarely GAG, which usually encodes valine), is found at the start of polypeptide chains, whereas UAA, UAG and UGA do not specify amino acids, but indicate termination or stop signals at the end of chains.

POSSIBLE CODONS OF AN AMINO ACID

First position (5' end)	Second position				Third position (3' end)
	U	C	A	G	
U	Phe (UUU)	Ser (UCU)	Tyr (UAU)	Cys (UGU)	U
	Phe (UUC)	Ser (UCC)	Tyr (UAC)	Cys (UGC)	C
	Leu (UUA)	Ser (UCA)	Stop (UAA)	Stop (UGA)	A
	Leu (UUG)	Ser (UCG)	Stop (UAG)	Trp (UGG)	G
C	Leu (CUU)	Pro (CCU)	His (CAU)	Arg (CGU)	U
	Leu (CUC)	Pro (CCC)	His (CAC)	Arg (CGC)	C
	Leu (CUA)	Pro (CCA)	Gln (CAA)	Arg (CGA)	A
	Leu (CUG)	Pro (CCG)	Gln (CAG)	Arg (CGG)	G
A	Ile (AUU)	Thr (ACU)	Asn (AAU)	Ser (AGU)	U
	Ile (AUC)	Thr (ACC)	Asn (AAC)	Ser (AGC)	C
	Ile (AUA)	Thr (ACA)	Lys (AAA)	Arg (AGA)	A
	Met (AUG)	Thr (ACG)	Lys (AAG)	Arg (AGG)	G
G	Val (GUU)	Ala (GCU)	Asp (GAU)	Gly (GGU)	U
	Val (GUC)	Ala (GCC)	Asp (GAC)	Gly (GGC)	C
	Val (GUA)	Ala (GCA)	Glu (GAA)	Gly (GGA)	A
	Val (GUG)	Ala (GCG)	Glu (GAG)	Gly (GGG)	G

The genetic code

The 20 amino acids [alanine (Ala), arginine (Arg), asparagine (Asn), aspartic acid (Asp), cysteine (Cys), glutamine (Gln), glutamic acid (Glu), glycine (Gly), histidine (His), isoleucine (Ile), leucine (Leu), lysine (Lys), methionine (Met), phenylalanine (Phe), proline (Pro), serine (Ser), threonine (Thr), tryptophan (Trp), tyrosine (Tyr) and valine (Val)] are encoded by base triplets. UAA, UAG and UGA cause termination of transcription.

GENE TRANSCRIPTION: TRANSMITTING THE CODE

RNA transcription can be divided into stages of:
* initiation;
* elongation;
* termination.

As with DNA replication, the process of gene transcription is more fully understood in prokaryotic than eukaryotic organisms. This in part reflects the simpler nature of prokaryotic genomes, which consist of closely packed genes, in which the coding DNA sequences are rarely interrupted.

Prokaryotic gene transcription

Gene transcription in prokaryotic cells (Fig. 1.10) is relatively simple.
* *Initiation* — RNA Pol binds to a specific sequence, known as a promoter which lies just upstream of the coding sequence of the gene.
* *Elongation* — RNA Pol then proceeds through the gene, synthesizing a RNA chain which is a copy of the antisense strand of DNA. The reaction is similar to the polymerization of DNA (see p. 17), in that nucleotide triphosphates are hydrolyzed releasing pyrophosphate which is split into inorganic phosphate.
* *Termination* — when the polymerase encounters a specific, known as a terminator, transcription stops and the completed RNA molecule is released.

TRANSCRIPTION IN PROKARYOTES

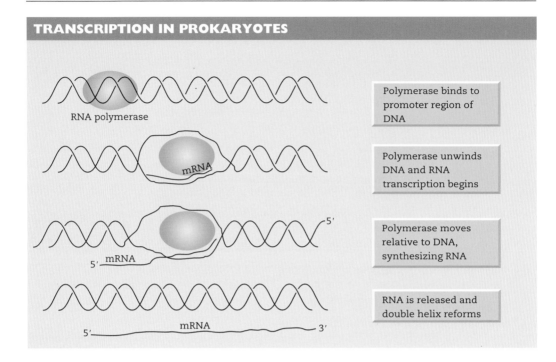

RNA polymerase

Polymerase binds to promoter region of DNA

mRNA

Polymerase unwinds DNA and RNA transcription begins

5′ mRNA

5′

Polymerase moves relative to DNA, synthesizing RNA

RNA is released and double helix reforms

5′ ———— mRNA ———— 3′

Fig. 1.10 Transcription in prokaryotes.

Gene structure and transcription in eukaryotic cells

Differences in both gene structure and the RNA Pol molecules make transcription in eukaryotes more complex.

In eukaryotic cells:
• there are three different types of RNA Pol;
• transcription is regulated by numerous different proteins, most of which bind to specific sites around the coding region of the gene;
• RNA is modified in several ways before being ready for translation into protein.

Only a small fraction of the human genome encodes polypeptides; over 95% is non-coding with no known function. Non-coding DNA occurs both within and between genes. Some of this apparently functionless DNA contains repetitive sequences of bases, which may be of functional significance. For example, centromeres which ensure complete disjunction at the middle of chromosomes, and telomeres which allow complete replication at the ends of chromosomes, both contain arrays of tandemly repeated DNA.

TRANSCRIPTION UNIT

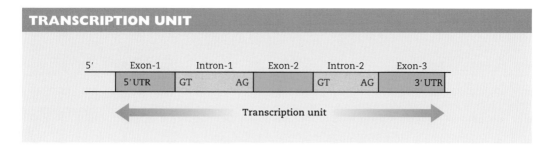

Fig. 1.11 Diagrammatic representation of a eukaryotic transcription unit. UTR, untranslated region.

CENTROMERES, TELOMERES AND ARRAYS

The *centromere* is the site at which chromosomes constrict during metaphase. It separates the long and short arms of the chromosome.

The *telomere* forms the end of the chromosome.

In *tandemly repeated arrays* identical DNA sequences appear one after the other along a stretch of DNA.

The structure of eukaryotic genes is more complex than a series of codons. The transcriptional unit of a gene is the region transcribed into a primary RNA transcript, which is a precursor of mRNA. It is made up of *exons* (containing expressed or coding DNA), which are interrupted by sequences of unknown function known as *inter*vening sequences (IVS) or *introns* (Fig. 1.11).

Introns begin with GT and end with AG.

The coding regions in the first and last exons are flanked by untranslated regions (UTR), which are actually part of the exons and are transcribed into mRNA but not translated into protein.

The rate at which genes are transcribed is regulated by DNA sequences, that usually lie outside the transcription unit, but on the same DNA strand. They tend to occur close to the gene, but may be located thousands of base pairs away.

Sequences on the 5' side of a region of DNA (to the left in a 5'–3' sequence) are often called 'upstream', whereas those on the 3' side are 'downstream'.

DNA sequences that influence the transcription of genes are known as *cis-acting control elements*. They do not encode proteins, but often influence gene transcription by acting as binding sites for protein products of other genes, known as *trans-acting transcription factors* or *DNA-binding proteins*. The mechanism by which such DNA-binding proteins influence

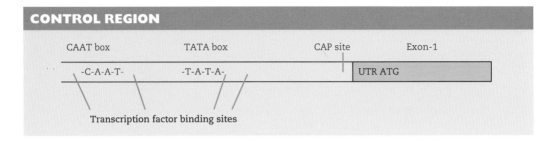

CONTROL REGION

CAAT box   TATA box   CAP site   Exon-1

-C-A-A-T-   -T-A-T-A-   UTR ATG

Transcription factor binding sites

Fig. 1.12 The promotor region of a gene.

gene transcription is not fully understood. It is possible that unfolding or folding of DNA to expose or hide certain sequences, or changes in the position of DNA with respect to the nuclear membrane, may be important.

cis-acting transcription factors are usually organized in clusters, which are located in the *promoter* and *enhancer* regions of the gene.

TRANSCRIPTION AND CONTROL REGIONS

The promoter region is located immediately upstream of the gene coding region and contains sequences that govern the rate of transcription, and define the site at which it starts.

Enhancer/silencer regions may be located within, near or some distance away from the gene whose expression they stimulate, or sometimes suppress.

Structures which favour the binding of proteins to DNA have been identified in some DNA-binding proteins. For example, *zinc fingers* consist of a fold of about 30 amino acids around a zinc atom, that seems to insert into grooves in the DNA helix. *Leucine zippers* contain four or five leucine residues spaced exactly seven residues apart.

Certain DNA sequences are found in the *promoter region* of most genes (Fig. 1.12). These include the TATA and CAAT boxes.

• *TATA box*—consists of an AT-rich sequence (often TATAA) which occurs about 30 bp upstream from the transcriptional start site (often denoted −30: the position of the nucleotide at the start site is designated +1). TATA boxes are often absent from the promoters of 'housekeeping genes'. Housekeeping genes, such as genes encoding the structural protein actin, are continuously expressed at low levels, and often have GC-rich

TERMINATION OF TRANSCRIPTION

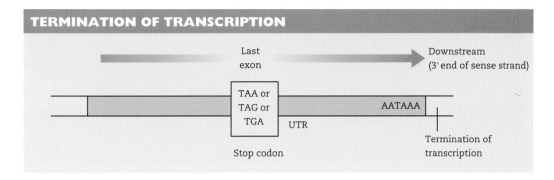

Fig. 1.13 Termination of eukaryotic transcription.

sequences such as GGGCGG in their promoter regions.
• *CAAT box*—contains this short sequence about 80 bp upstream (−80) of the start site.

These sequences, together with binding sites for other transcription factors which vary according to the gene involved, are responsible for the rate of transcription.

Transcription starts at the *CAP site*, so-called because following transcription, the 5′ end of the mRNA is capped at this site by the attachment of a specialized nucleotide (7-methyl guanosine). The cap site is followed by a *leader sequence* leading up to the *initiation codon* (ATG), that specifies the start of translation.

Transcription then proceeds, such that a full copy of the gene (*introns* and *exons*) is made using a process similar to DNA replication (Fig. 1.14). A *stop codon* (TAA, TAG or TGA) indicates the end of the translated region (Fig. 1.13). This is followed by an UTR, which includes the *poly(A) signal* (AATAAA), that signals cleavage of the newly formed RNA at a position slightly downstream, and the addition of a string of adenosine residues (a poly(A) tail). The poly(A) tail is thus not encoded by the gene, rather adenosine residues are sequentially added enzymatically following transcription.

Thus, each end of the mRNA contains an UTR which is not translated into protein. The function of these regions is not fully understood, although the 5′ region (upstream) appears to influence translation, whereas the 3′ region (downstream) may contain sequences that are important in the stability of mRNA.

The 5′ cap binds to the small subunit of the ribosome as the first step in translation.

In eukaryotic cells, transcription is mostly controlled at the level of initiation. The binding of transcription factors to the promoter region of a gene attracts RNA Pol.

EUKARYOTIC TRANSCRIPTION

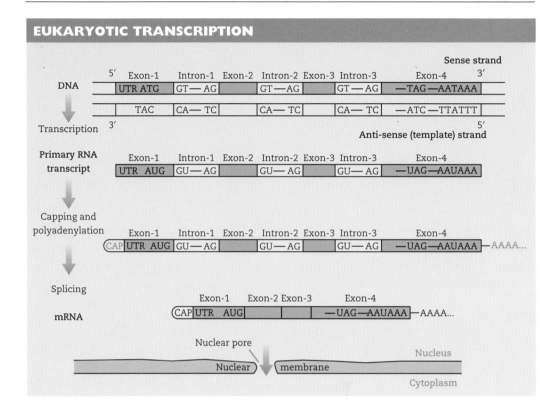

Fig. 1.14 Transcription in a eukaryotic cell.

The primary RNA transcripts produced by RNA Pol II are referred to as *heterogeneous nuclear RNA (hnRNA)*, because unlike tRNA and rRNA, they show considerable variability (heterogeneity) in size.

Most rRNA is transcribed by RNA Pol I in the nucleolus, whereas the small (5S) rRNA and tRNA are transcribed by RNA Pol III in the extranucleolar region.

Three types of RNA Pol (I, II and III) occur in eukaryotes, of which RNA Pol II is principally involved in transcription of mRNA.

The presence of four different proteins which act as transcription factors is required before a RNA Pol II molecule can recognize and bind to the promoter of a gene and start transcription. These are referred to as TFIIA, TFIIB, TFIIC and TFIID because they each act as *transcription factors* for RNA Pol II. TFIID binds to the TATA box and is also known as the TATA factor.

Binding of RNA Pol II to the promoter region of a gene results in a localized separation of double-stranded DNA. During formation of a mRNA molecule about 20 bp of DNA are unwound at any one time, of which around 10 form a DNA:RNA hybrid. The RNA Pol proceeds through the gene and synthesizes a mRNA chain from 5′ to 3′ end by adding ribonucleoside triphosphate bases complementary to the DNA. Only one strand (the anti-sense strand) is used as a

ALTERNATIVE SPLICING

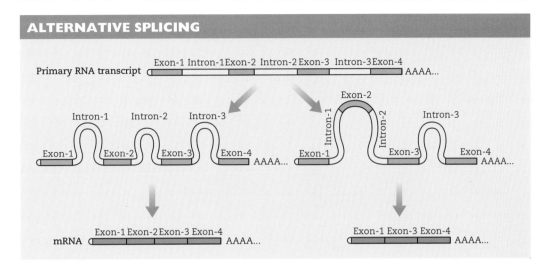

Fig. 1.15 Alternative splicing produces different mRNAs from the same primary transcript.

template. The primary RNA transcript is thus an exact copy of the sense DNA strand, except that U replaces T.

Post-transcriptional processing

Whilst still in the nucleus the newly synthesized RNA is modified by the following events.
• *Capping*—the addition of a nucleotide cap.
• *Polyadenylation*—detachment of the RNA and addition of a string of adenosine residues.
• *Splicing*—sequences corresponding to *introns* are *excised* and discarded, and the remaining *exons* are *spliced together*.

During splicing all of the introns are usually removed, leaving all of the exons in the mRNA. However, exons may also be removed during the splicing process, resulting in variations in the final mRNA product, and hence, the polypeptide it encodes. The process by which different mRNA transcripts are formed by removal of different segments of the primary RNA transcript is known as alternative splicing (Fig. 1.15).

After capping, polyadenylation and splicing, the RNA is then ready for transport to the cytoplasm.

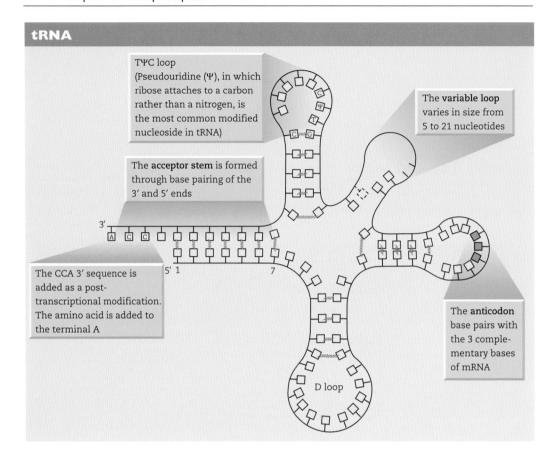

Fig. 1.16 The structure of tRNA.

TRANSLATION: READING THE CODE

Protein synthesis occurs on ribosomes. In eukaryotes, ribosomes consist of 40S (small) and 60S (large) subunits, which together form 80S particles (S = Svedberg units, p. 16). The 60S subunit contains a polypeptide complexed to three rRNAs (28S, 5.8S and 5S), whereas in the 40S subunit the polypeptide is complexed to 18S RNA.

The process of protein synthesis is started by the formation of a complex involving a 40S ribosomal subunit carrying a methionine tRNA, which base pairs with the initiation codon AUG on mRNA (Fig. 1.16).

Once the initiation complex has formed, synthesis of the polypeptide chain is driven by elongation factors (elFs), which join the 60S subunit to the complex and move the ribosome relative to the mRNA (Fig. 1.17). At each codon they promote the formation of peptide bonds between amino acids which are

TRANSLATION: MAKING PROTEINS

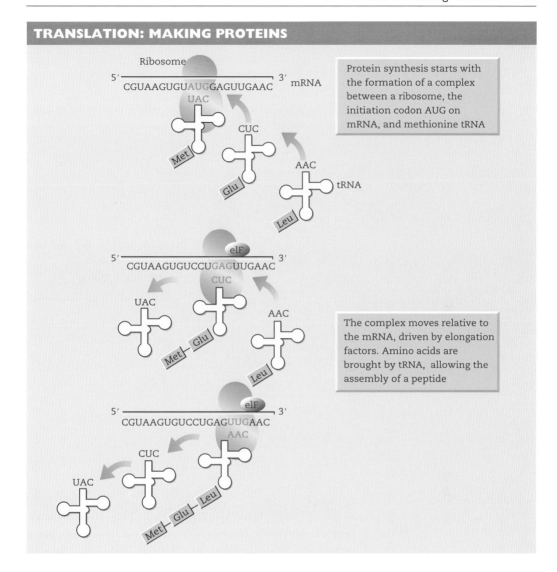

Protein synthesis starts with the formation of a complex between a ribosome, the initiation codon AUG on mRNA, and methionine tRNA

The complex moves relative to the mRNA, driven by elongation factors. Amino acids are brought by tRNA, allowing the assembly of a peptide

Fig. 1.17 Translating the genetic code of mRNA into the polypeptide chain of a protein.

delivered by tRNA. Each rRNA carries an amino acid and a triplet of bases (*anticodon*), which recognize a codon on mRNA specific for the amino acid. For example, the tRNA that carries methionine has the anticodon UAC which recognizes the methionine codon AUG on mRNA. When the ribosome reaches a termination codon (UAA, UAG or UGA) the completed polypeptide is released from the last tRNA, and the ribosomal units fall off the mRNA.

DIAGRAM OF A CELL

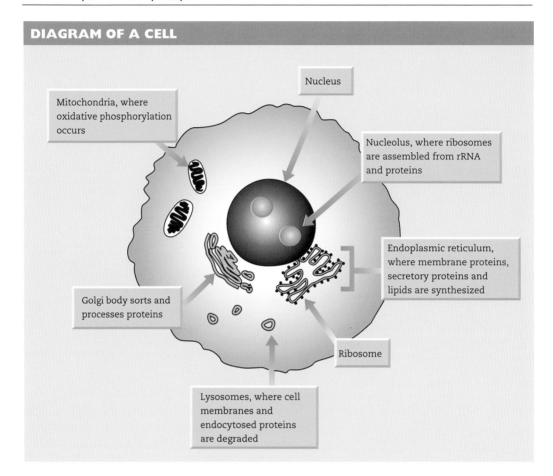

Nucleus

Mitochondria, where oxidative phosphorylation occurs

Nucleolus, where ribosomes are assembled from rRNA and proteins

Endoplasmic reticulum, where membrane proteins, secretory proteins and lipids are synthesized

Golgi body sorts and processes proteins

Ribosome

Lysosomes, where cell membranes and endocytosed proteins are degraded

Fig. 1.18 Diagrammatic representation of a eukaryotic cell showing some of the important organelles.

Post-translational processing

Polypeptides may start to form the complex structure of proteins as they are synthesized. Secretory proteins pass through the rough endoplasmic reticulum and move to the Golgi body for processing (Fig. 1.18). Following synthesis many polypeptides are modified further, e.g. by hydroxylation or phosphorylation of amino acids, or addition of sugars (glycosylation).

CHAPTER 2

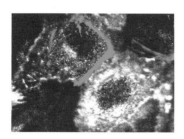

Manipulation of DNA/RNA

Molecular medicine is based on the ability to manipulate biomolecules. The next two chapters are intended to provide an overview of the techniques involved. Rather than providing specific protocols, the emphasis is on understanding some of the fundamental principles that underlie the rapidly evolving protocols. The basic techniques and important enzymes are introduced in this chapter, the next chapter shows how these tools are used to answer specific questions. Such background should make one a more informed and critical reader of many current research papers, and help in appreciating the possibilities and identifying the limits of gene technology.

PREPARATION OF NUCLEIC ACID

A cell is a mixture of relatively simple nucleic acids and complex everything else. Indeed, their relative structural simplicity and homogeneity led many early researchers to doubt that nucleic acids *could* be the 'stuff' of which genes are made. These researchers thought that nucleic acids were probably only structural elements of chromosomes, perhaps organizing the proteins that were deemed complicated enough to constitute genes. The fact that genes are composed of only four biochemically similar nucleotides that convey information in their sequence without requisite tertiary structure, makes the chemistry of isolation and characterization correspondingly uniform, i.e. independent of their sequence.

Nucleic acids are uniformly and strongly negatively charged because of their phosphate backbones and, therefore, they prefer an aqueous environment where these charges are solvated. Everything else (proteins, lipids, carbohydrates) contains charged and uncharged regions as well as hydrophobic and hydrophilic regions, therefore preferring either a hydrophobic (organic) environment or the interface between organic and aqueous phases. This is the basis of extraction by the organic solvent phenol and other separations. Nucleic acids are also relatively dense and this is the basis of separation on caesium chloride (CsCl) gradients.

First, however, we need to introduce the broad groupings of genetic material and their general physical characteristics before embarking on a discussion of their manipulation.

DNA preparation

DNA is easy to prepare and store, principally because DNA-specific degrading enzymes (DNases) are easily destroyed by moderate heating (65°C) or inhibited by compounds such as ethylenediaminetetra-acetic acid (EDTA) that sequester (chelate) divalent ions. For some purposes, e.g. preparing a template for the polymerase chain reaction (PCR), it is often sufficient simply to lyse the cells (i.e. destroy the cells through damage or rupture of the plasma membrane) and denature the proteins and DNA at high temperature (95°C).

DENATURATION

Biological molecules have a natural configuration that is necessary for their activity. When this native configuration is disturbed or destroyed, the molecule is said to be *denatured*. For proteins, the configuration is determined by a relatively complex interaction of structural elements. For DNA, the principal structural element is its double-stranded nature. Thus, DNA is said to be denatured when it is made single stranded and renatured when the double strand is reformed. RNA is made as a single strand that then folds to form double-stranded regions. RNA is said to be denatured when these regions are made single stranded.

When native (double-stranded) DNA or purer preparations are required, protein can be extracted by phenol or separated by centrifugation through a density gradient formed by CsCl solutions.

Chromosomal DNA: really big and stringy

Chromosomal (genomic) DNA is difficult to prepare without shearing (tearing apart) at least some of the DNA. This is hardly surprising when one considers that the DNA contained in one mammalian cell is nearly 2 m long (stretched end-to-end and deprived of packaging proteins), so the DNA from even a few cells can produce quite a knot. Even the much shorter chromosomal DNA from bacteria ($\sim$1 mm) can quickly become a mess.

Fortunately, it is rare that very large pieces of DNA are analysed, so the practical problem is more that chromosomal DNA is extremely difficult to pipette (transfer from one tube to another) because it is so stringy (viscous). When large pieces of DNA ($>$50 kbp) are needed intact, as when mapping genes by pulsed-field gel electrophoresis (PFGE, p. 46), cells can be lysed, deproteinated and the DNA digested *in situ*, right in the gel.

Plasmid DNA: pretty small and circular

Plasmids are small, circular DNAs that are extrachromosomal and independently replicating in bacteria, i.e. they have their own origin of replication (*ori*). Bacteria have been using plasmids forever to exchange useful genes; molecular biologists have caught on only in the last couple of decades. Plasmids are quite easy to isolate from bacteria, manipulate *in vitro* and reintroduce into bacteria. Therefore, they are often used as *cloning vectors*, meaning that DNA is inserted into the plasmid and 'grown-up' along with the rest of the plasmid DNA.

Recombinant DNA techniques have been used to improve some plasmids for use as cloning vectors in the following ways (Fig. 2.1).

• By adding restriction enzyme sites that allow the plasmid to be cut at particular sites (e.g. the polylinker).

• By removing DNA to make the plasmid smaller and to delete redundant restriction enzyme sites (e.g. the *Eco*RI site at position 0), which makes the plasmid easier to manipulate.

• By adding genes that allow simpler screening and selection procedures to find successful recombinants (e.g. *lacZ*).

• By substituting a mutant *ori* that allows *relaxed copy number* control, meaning that the plasmid continues to replicate after the bacterium has stopped. This may not be desirable for

IMPORTANT PARTS OF A PLASMID

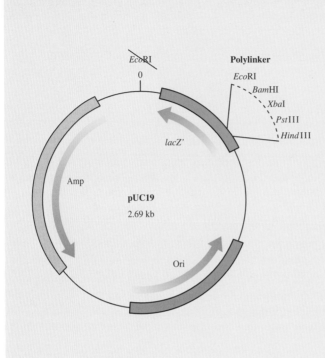

Polylinker Also called the multiple cloning site (MCS), it contains many useful restriction enzyme sites (see below) for inserting exactly that piece of DNA that you want in this plasmid

Amp Gene that confers resistance to the antibiotic ampicillin, permitting selection for bacteria that contain the plasmid (see selection)

lacZ' Gene that encodes a subunit of the easily assayed enzyme β-galactosidase, permitting rapid screening for recombinant plasmids (see screening)

Ori The **ori**gin of replication regulates the number of plasmid copies per bacterium

Fig. 2.1 Map of the important parts of a plasmid, the workhorse of molecular biology. Shown is the plasmid pUC19, a late-generation cloning vector of 2686 bp (2.69 Kb) that was constructed from parts of previous plasmids using recombinant DNA technology. The '0' point at the top is where the numbering of the base pairs starts, proceeding clockwise. An EcoRI restriction enzyme site that was originally at this position was deleted during the construction, as indicated by the line through the name, to maintain the uniqueness of the EcoRI site in the polylinker.

CIRCULAR NATURE OF PLASMIDS

The small circular nature of plasmids is exploited in their purification. Plasmid DNA is often liberated from the bacterium under denaturing conditions, which causes the DNA strands to come apart. Upon return to conditions favouring renaturation, the complex bacterial chromosomal DNA cannot find the other matching strand and instead forms short, weak, double-stranded regions. The strands of circular plasmid DNA, in contrast, remain linked and efficiently reanneal.

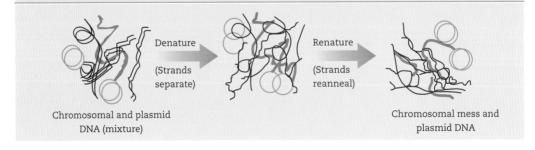

Chromosomal and plasmid DNA (mixture)

Denature

(Strands separate)

Renature

(Strands reanneal)

Chromosomal mess and plasmid DNA

RNA AND ITS PREPARATION

RNA name (aliases)	What is it?	How much is there?	How is it made?
Total	All the RNA in the cell	~1% of cell mass	Guanidinium lysis, then phenol extraction or CsCl ultracentrifugation
Cytoplasmic	Non-nuclear	Less than total but still a lot	Lysis with low detergent concentrations, which keeps nucleus intact, purify RNA from cytoplasm
Messenger (mRNA) (poly (A+))	RNA possessing a 'tail' of A nucleotides	Very little (3% total). *Naturally, the most desirable*	Prepare total or cytoplasmic RNA, purify poly(A+) by binding ('annealing') to poly(T), then wash and elute by denaturing the RNA–DNA hybrids. *Extremely heterogeneous in size*
Ribosomal (rRNA)	Structural components of ribosomes	Most of the total RNA	Only rarely Four sizes: 28S (4718bp), 18S (1874bp), 5.8S (160bp) and 5S (120bp)
Transfer (tRNA)	The adapters between mRNA and amino acids	Plenty	Only rarely About 100 different forms, 75–85bp

Table 2.1 RNA forms

the bacterium or the symbiotic plasmid, but it is great for producing a much higher yield of DNA for the molecular biologist.

Plasmid DNA may be *incorporated* into genomic DNA following transfection (transfer) into mammalian cells, which produces a stable transfectant. There are also plasmids that can replicate autonomously in mammalian cells; these plasmids include pieces of animal viruses such as SV40 (Simian virus 40) or retroviruses. Yeast and mammals also have natural circular, independently replicating, extrachromosomal DNAs (e.g. mitochondrial DNA), but they are larger, do not possess multiple cloning sites, screening or selection functions, and they do not replicate in bacteria, so they lack many of the attributes of a good cloning vector.

cDNA (see RNA)

The 'c' in cDNA stands for *complementary*, which means that cDNA is made to complement messenger RNA (mRNA). So, you first have to prepare RNA and then reverse transcribe it into cDNA (see below).

RNA preparation Table 2.1

RNA preparation is difficult because RNases (enzymes that

degrade RNA) are extremely stable and can even refold and regain activity after complete denaturation. Therefore, methods of isolating RNA attempt to:

1 inhibit cellular RNases;
2 separate RNA from DNA and proteins (including RNases);
3 keep RNase out.

Many protocols employ lysis buffers containing guanidinium, which is a strong denaturing agent and a strong inhibitor of RNases. The problem with guanidinium is that it also inhibits nearly everything else. Therefore, when the experiment calls for subsequent enzymatic manipulation of the RNA, an additional ethanol precipitation of the RNA (see below) is often used to reduce the amount of residual guanidinium.

Aside from the cells themselves, fingers are an important source of RNases. Therefore, extra care is taken in preparing solutions, tubes and pipette tips. Wearing gloves may give a false sense of security since they can quickly become contaminated with RNase after handling pipettes, bottles, etc. Tubes and pipette tips are often autoclaved or baked, treatments that can destroy some of the RNase. Some solutions can be treated with diethylpyrocarbonate (DEPC), which irreversibly denatures RNases, and then autoclaved to destroy the DEPC. Many solutions cannot be treated with DEPC, however, because it also reacts with the solute. Also, residual DEPC can ruin experiments, so instead it may be better to use caution and prepare the purest solutions possible. RNases can be also specifically inhibited with RNasin (*RNase inhibitor*), a labile protein that is relatively expensive, but is also effective and safe.

Although ribosomal RNA (rRNA) is rarely specifically prepared, its presence in most RNA preparations is useful. It acts as a carrier for the rarer mRNA and it produces strong, easily detectable bands when total or cytoplasmic RNA preparations are analysed by gel electrophoresis, which separates molecules by size. The heterogeneously sized mRNA produces only a relatively faint smear on gels. The sharpness and relative intensities of two main rRNA bands are also good indicators of the mRNA quality and quantity. Mammalian RNA produces two prominent bands called 28S and 18S of lengths 4718 and 1874 bp, respectively. (S is for Svedberg, the unit of measurement for speed of sedimentation upon ultracentrifugation.) If these bands are sharp and the 28S band is about twice as bright as the 18S band, then the RNA preparation has probably not suffered much digestion by RNases. Although the smaller 5.8S (160 bp) and 5S (120 bp) rRNAs are present in equal numbers (they are all subunits of the ribosome), their small size makes them relatively difficult to detect on gels.

Common techniques in nucleic acid preparation and analysis

Optical density (OD) measurement: how much, how pure?

Nucleotides in solution absorb light in the ultraviolet (UV) region of the spectrum, with a maximum absorption around a wavelength of 260 nm (Table 2.2). Proteins also absorb UV light, but more at the 280 nm wavelength and less at 260 nm. Therefore, nucleic acids can be quantified and the purity of a preparation can be estimated by measuring the absorption at both 260 and 280 nm (spectrophotometry) and comparing the results to those obtained from pure solutions. The term OD is commonly used instead of absorbance.

The OD at additional wavelengths may be measured to determine the extent of contamination by chemicals used in the preparation. For example, phenol contamination can be estimated from the OD 300, whereas guanidinium contamination can be estimated at OD 230, although many other compounds also absorb at these wavelengths (Tables 2.3 and 2.4).

OD OF NUCLEOTIDES

	Deoxynucleotides			
	dATP	dTTP	dGTP	dCTP
OD 260*	1.52	0.84	1.20	0.71

* Shorthand for 'The OD at a wavelength of 260 nm (UV light) of a 0.1 mM/l solution of this deoxynucleotide is . . .'

Table 2.2 Nucleotides absorb UV light with a maximum at 260 nm. For each nucleotide, the OD at 260 nm of a 0.1 mM/l (100 μmol/l) solution is listed. These values are useful for determining the concentrations of nucleotides or shorter oligonucleotides. Note that the absorbances of the two pyrimidines cytosine (C) and thymine (T) and the two purines adenine (A) and guanine (G) are similar to each other, so differences in nucleotide composition (percentage of G/C) have less effect on OD

Table 2.3 OD measurement at different wavelengths – finding out what is in the solution

NUCLEIC ACID AND CONTAMINANTS

Measure what?	Where?
Nucleic acid (mostly)	260 nm
Protein (mostly)	280 nm
Nucleic acid purity	Ratio 260/280
Guanidinium (etc.)	230 nm
Phenol (etc.)	300 nm

ODs OF PURE SOLUTIONS

Pure solution	OD 260 of 1=	Ratio OD 260/280
DNA	50 μg/ml	1.8
RNA	40 μg/ml	2.0

Table 2.4 Observed ODs and ratios of pure solutions. These values are used to determine the amount and purity of a DNA or RNA preparation. If, for example, protein is present in the solution, the OD 280 increases more than the OD 260 (because proteins absorb light of 280 nm wavelength better than 260 nm), so the ratio OD 260/OD 280 decreases

HYPOCHROMICITY

A couple of quick calculations would tell you that there is either some extra DNA or some missing OD in the observed OD 260 values, because these values are lower than the sum of the absorbances of the constituent nucleotides.

Hypochromicity: *or, the whole absorbs less than the sum of its parts.*

An average nucleotide = (1.52 + 0.84 + 1.20 + 0.71)
    absorbs   ÷ four nucleotides (see Table 2.2)
                 = 1.07 OD 260/0.1 mM nucleotide.

So, a solution with   = 1 ÷ 1.07 OD 260/0.1 mM nucleotide
    an OD 260 of 1   = 0.093 mM nucleotide (*calculated*).

But, a solution of DNA with an OD 260 of 1 = 50 μg/ml (see Table 2.4) which, when converted to yield the molarity of nucleotides . . .
50 μg/ml × 1 μmol/330 μg (nucleotide molecular weight)
                 = 0.152 μmol/ml
                 = 0.152 mM nucleotide (*observed*).

This is probably because the nucleotides are stacked up in double-stranded DNA, placing many nucleotides in their neighbours' shadow where they cannot absorb all the UV light that they otherwise would if they were still single. Therefore, one should calculate the amount of purified double-stranded DNA using the observed values (see Table 2.4) and the amount of nucleotide or shorter oligonucleotides using the ODs of the individual nucleotides (see Table 2.2).

Ethanol precipitation: concentration or buffer change

Even small amounts of DNA or RNA can be efficiently precipitated (made insoluble) by adding salt and alcohol and centrifuging. The addition of salt increases the ionic strength of the solution, which reduces the repulsion of the like- (negatively) charged phosphate groups on the nucleic acid backbone, thereby allowing the nucleic acid molecules to come closer together. The addition of alcohol makes the solution more hydrophobic and therefore less able to solvate the charged nucleic acid. These effects combine to reduce the solubility of the nucleic acids and produce the precipitate.

DNA/RNA PRECIPITATION

DNA or RNA
+ salt (to taste*)
+ alcohol†
= precipitated nucleic acid (spin hard‡ to collect)
Ethanol precipitation of nucleic acids. *Sodium chloride or sodium acetate (0.2 mM), ammonium acetate (2.5 mM) or lithium chloride (0.8 mM) are often used. †2 volumes ethanol for DNA, 2.5 volumes ethanol for RNA or 1 volume 2-propanol (isopropanol). ‡Centrifuge 'microfuge' 10 minutes at $\sim 12000 \times g$.

The early protocols called for the mixture to be incubated in the cold and many 'old-timers' still keep their ethanol in the freezer and place their tubes into dry ice/ethanol baths. Careful analysis later found freezing to be less efficient than room temperature incubation for standard DNA precipitation. The precipitate is collected by centrifugation and the pellet is then dried and resuspended in the desired buffer. Several different salts can be used; the choice of salt is dictated by the presence of additional compounds in the nucleic acid preparation or the intended use of the nucleic acid. For example, ammonium acetate precipitation after a labelling reaction (which adds traceable nucleotides to the nucleic acid) efficiently removes unincorporated nucleotides. However, the residual salt can inhibit some enzymes (e.g. polymerases, which add nucleotides, or kinases, which add phosphate) that may be called for in subsequent reactions.

Ethidium bromide (EtBr): fluorescent stain of DNA and RNA

EtBr is a small molecule that inserts (intercalates) between the nucleotides of DNA or RNA and strongly fluoresces under UV illumination. This forms the basis of a quick, easy and sensitive detection system or 'stain' of nucleic acids. When EtBr binds to DNA or RNA, it is effectively concentrated, so the nucleic acid shows up as a bright band amidst a dimmer background fluorescence of unbound EtBr.

CsCl/ultracentrifugation: purification

RNA is denser than DNA, which in turn is denser than the other cellular material that you do not want (protein, lipids, polysaccharides, etc.). So, whilst RNA sinks below a solution of CsCl of a certain density, DNA stays in the CsCl and the other material floats above the CsCl solution (Fig. 2.2). This separ-

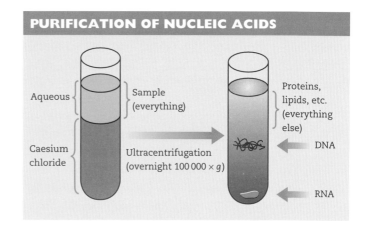

Fig. 2.2 CsCl purification of nucleic acids. The EtBr-stained nucleic acids fluoresce under UV illumination and would actually appear as bright bands.

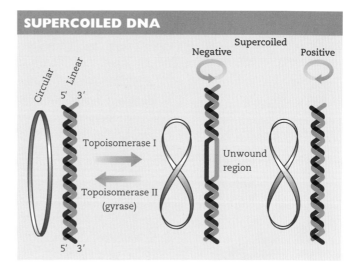

Fig. 2.3 Supercoiled is the natural way. The supercoiled DNA forms a denser structure than separate strands. Note that in circular DNA, rotation of strands relative to each other *must* be some integral number of turns, because half turns would juxtapose unligatable ends (3'–3' or 5'–5').

ation occurs during ultracentrifugation ($100\,000 \times g$) in the presence of EtBr.

A special use of CsCl/ultracentrifugation is to purify plasmids. Most chromosomal and plasmid DNA is naturally negatively supercoiled, meaning that they possess an extra twist counter-clockwise, once every ~200 bp (Fig. 2.3). The maintenance of the supercoiled state in plamids is dependent upon the integrity of both DNA strands—a break in one strand (a 'nick') allows the unwinding of the supercoil and the DNA molecule reverts to the lower energy, relaxed form. A single-stranded break results in the covalently closed circular form, whereas juxtaposed double-stranded breaks result in the linear form. Supercoiled DNA molecules are more dense than relaxed circular or linear molecules, so they 'float' lower on the CsCl density gradient.

CsCl density purification in the presence of EtBr complicates the mechanism, but does not change the result. When EtBr intercalates between the nucleotides, it pushes the nucleotides further apart and unwinds the helix, first unwinding the negative supercoils and then inducing positive supercoils (clockwise twists). In a closed (supercoiled) circle there is only a limited amount of intercalation and unwinding that can occur, because the ends of the molecule are fixed. Therefore, the supercoiled plasmids allow only a limited amount of EtBr to intercalate, remain denser and float relatively lower on the CsCl gradient than the chromosomal DNA.

Phenol extraction: protein removal

Phenol is an organic solvent that is often used to separate nucleic acids and proteins. Phenol is added to the mixture of proteins and nucleic acids, the mixture is shaken (usually vortexed unless large-molecular-weight DNA is being prepared, in which case it is shaken gently to reduce shearing of the DNA), and then the tube is centrifuged to accelerate the separation of the organic and aqueous phases (Fig. 2.4). Proteins are denatured and segregate into the phenol or remain at the aqueous interface. Nucleic acids stay in the aqueous phase at near-neutral pH because of their highly negatively charged phosphate backbone.

Phenol is toxic, and a notable alternative for the removal of proteins is to make proteins insoluble with high salt concentrations and then to remove the protein precipitate by centrifugation. Because salt, unlike phenol, is not a denaturant,

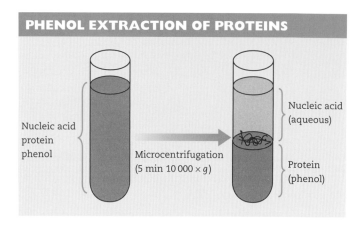

PHENOL EXTRACTION OF PROTEINS

Nucleic acid
protein
phenol

Microcentrifugation
(5 min 10 000 × *g*)

Nucleic acid
(aqueous)

Protein
(phenol)

Fig. 2.4 Phenol extraction of proteins from nucleic acids.

pH DEPENDENT PARTITIONING OF NUCLEIC ACIDS

Phenol pH	Aqueous phase	Organic phase
Neutral (pH 7–8)	RNA+DNA	Protein, etc.

Table 2.5 Acidic phenol is used to prepare RNA depleted of DNA

high salt precipitation may not be quite as effective as phenol in removing proteins. However, nucleic acids can be immediately and selectively precipitated out of the protein-depleted high salt solution by the addition of ethanol.

An increasingly popular protocol for extraction of RNA uses acidic phenol, which keeps the RNA in the aqueous phase but sends the DNA into the phenol phase, probably because the phosphate groups on the DNA are more easily neutralized (i.e. less acidic) (Table 2.5).

Gels and columns (separation by size)

Gels and sizing columns separate molecules in opposite order: small molecules come out of a gel *first* and out of a column *last*. This is because a gel is a mesh that large molecules have a relatively hard time getting through, whereas columns contain beads with small pores that keep out large molecules, but allow in small molecules (Fig. 2.5). Since the small molecules can enter the beads, the volume of the column is bigger for them, whereas the excluded big molecules are left with only the volume between the beads, called the *void volume* because it is made up from the voids left between the beads that pack the column. So, the big molecules come out (elute) in a smaller volume (earlier).

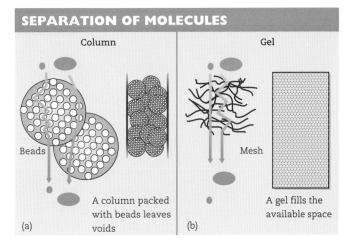

SEPARATION OF MOLECULES

Column

Gel

Beads

Mesh

(a)   A column packed with beads leaves voids

(b)   A gel fills the available space

Fig. 2.5 (a & b) Separation of molecules by size on sizing columns and gels. Note the reverse order of separation: bigger molecules pass more quickly through the column and more slowly through the gel.

The fluid in both columns and gels usually contains a buffer chosen to resist changes in pH. Molecules move through a column in the bulk flow of fluid, driven by gravity or by a pump. In contrast, molecules are driven through a gel by a voltage gradient. DNA and RNA are uniformly negatively charged and naturally adopt a stretched, rod-like configuration. This means that sieving of nucleic acids is a function of their length, unlike proteins that normally fold up compactly and have to be denatured with anionic detergent to be resolved according to polypeptide chain length.

METHODS OF COLUMN CHROMATOGRAPHY

Unlike sizing columns, which are also called molecular sieves, other methods of column chromatography separate molecules on the basis of characteristics other than size.

For example:

• ion-exchange chromatography separates molecules based upon their charge;

• affinity chromatography separates molecules based upon their ability specifically to bind the column material;

• reversed-phase chromatography separates molecules based upon their hydrophobicity.

Similarly, a form of gel electrophoresis called isoelectric focusing (IEF) separates molecules on the basis of their net charge. Although these are powerful analytical techniques, they are used much less often than sizing columns or gels.

Really big pieces of DNA, such as even a relatively small piece of a chromosome, cannot be separated by conventional gel electrophoresis. This is probably because the DNA begins to snake through the gel and the differences in the resistance to the movement of a short snake as compared to a long snake are too small to resolve. Pulsed field gel electrophoresis (PFGE) was developed to separate large pieces of DNA. The voltage gradient (field) is periodically reorientated so that large pieces of DNA cannot remain orientated 'end-on' and separation by size is possible.

Nucleic acids are often detected in the gel with EtBr (Fig. 2.6). This stain can detect <1 ng of DNA in a single band (the sensitivity with RNA probably depends upon the extent of double-strandedness). Safety hazards must be considered, however, because EtBr is mutagenic and toxic, so solutions and gels are handled with gloves and disposed of properly. Also, UV light can burn the retina and the skin (this is especially dangerous when DNA bands are cut out of preparative gels), so gloves and UV-blocking safety glasses are worn. As one might expect from these warnings, EtBr combined with UV light is also damaging to the DNA/RNA, so exposure should be minimized.

DNA is negatively charged at neutral pH so it migrates toward the positively charged electrode. The discrete size fragments are often called 'bands'. Note that the *amount* of DNA in a band determines the amount of EtBr bound and the resulting fluorescence, not the size (Fig. 2.6). Confusion on this point is perhaps a consequence of the typical experiment in which a particular DNA is digested and the resulting fragments are analysed by gel electrophoresis. The smaller and larger fragments are present in *equimolar* amounts (there are just as many molecules of any particular small fragment as there are

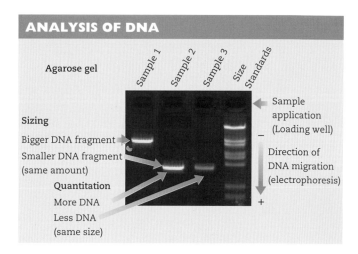

Fig. 2.6 Analysis of DNA by agarose gel electrophoresis allows the estimation of size and quantity. The DNA is detected by fluorescence with an EtBr 'stain'.

of any particular large fragment), but the larger fragments comprise *more mass* because they have more base pairs per molecule and, therefore, they fluoresce brighter. An example of such a digestion is in the right lane of Fig. 2.6, labelled 'standards'.

CUTTING AND PASTING DNA (RESTRICTION ENZYMES AND LIGASES)

The foundation of modern molecular biology is the ability to cut DNA at specific places, producing fragments that can be *recombined* in different patterns (recombinant DNA). Restriction endonucleases and ligases are the enzymes that perform these functions.

Restriction enzymes

Restriction endonucleases are enzymes that cut (digest) DNA at specific sequences (sites).

Restriction endonucleases with over 200 different sequence specificities have been characterized, so many different fragments can be generated. For example, the enzyme *Eco*RI (isolated from *Escherichia coli*) cuts DNA at the sequence GAATTC, whereas *Pst*I (isolated from a different bacterium) cuts at the sequence CTGCAG (Fig. 2.7).

Note the importance of the 5'-N-3' orientation as a convention for writing DNA sequences (Fig. 2.8). *Eco*RI, which cuts the sequence 5'-GAATTC, would not cut the same sequence of nucleotides in the opposite orientation, i.e. 3'-GAATTC. As it happens, the restriction enzyme *Afl*II would cut this sequence (5'-CTTAAG).

Over 2000 restriction enzymes have been isolated to date with around 200 different sequence specificities, which means that there are usually several enzymes that recognize a given

Fig. 2.7 Different restriction enzymes cut different sequences. 'N' means any DNA base and only the new terminal phosphate (p) is shown.

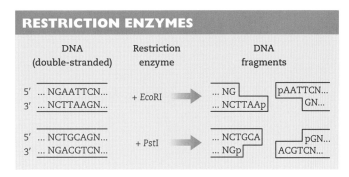

RESTRICTION ENZYMES

Almost all commonly used restriction enzymes cut rotationally symmetric sequences. The recognition sequence is often (wrongly) said to be mirror symmetric or palindromic as in the phrase, 'Madam, I'm Adam' (in which the comma/apostrophe is rotationally symmetric).

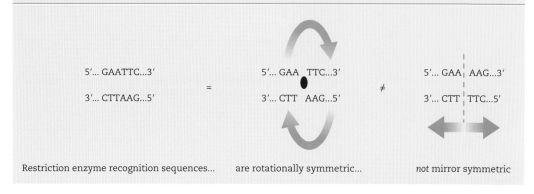

Restriction enzyme recognition sequences...   are rotationally symmetric...   *not* mirror symmetric

ORIENTATION

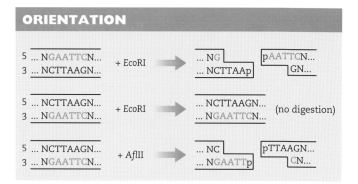

Fig. 2.8 Orientation is important because 5'-GAATTC is not the same as 3'-GAATTC.

ISOSCHIZOMERS

Isoschizomers   Products

Fig. 2.9 Isoschizomers are different restriction enzymes that recognize the same sequence but may cut in different ways.

sequence. When different restriction enzymes recognize the same sequence, they are called *isoschizomers*. They do not always cut at the same place in the sequence (Fig. 2.9).

The examples given above are all 6 bp recognition sites. The six-base cutters are the most commonly used because they tend to cut DNA into fragments that are small enough to handle yet are big enough to be useful. The average fragment length is a function of the size of the recognition sequence and the approximately random distribution of A/T/G/C in DNA (ignoring for now the fact that the mammalian genome is relatively AT-rich and CG-poor). With four bases to choose from, any given sequence of six bases will occur once every approximately 4000 bases ($4^6 = 4096$). In contrast, an 8 bp cutter like NotI is more useful for mapping larger pieces of DNA such as mammalian genomes, because it produces only 1/16th as many DNA fragments (of an average 64 000 bp).

The frequency of certain restriction enzyme sites is actually much lower than expected (Table 2.6). As noted above, mammalian genomes are A- and T-rich, which reduces the expected frequency of G- or C-containing restriction sites. Even so, the frequency of the CG dinucleotide is fivefold lower than would be expected based on G and C content. Consequently, the frequency of restriction enzyme sites containing CG is low in the mammalian genome. Furthermore, these CG sites are often methylated, which is a modification that blocks cutting by many restriction enzymes. Surprisingly, the genome is not homogeneous in this respect: stretches of several hundred to

GENOMES OF DIFFERENT ORGANISMS

DNA source	Genome size (kb)	Number of restriction sites		
		4 bp	6 bp	8 bp
1 pUC19	3	10	0–1	0–1
2 SV40	5	20	1	0–1
3 Bacteriophage λ	48	190	12	0–1
4 Bacteriophage T4	165	660	40	2–3
5 Bacteria	4700	18400	1100	70
6 Yeast*	16000	62500	3900	250
7 Fruit fly*	120000	470000	30000	1800
8 Mammals*	3000000	11700000	730000	46000

* Haploid values (most somatic cells have twice as much DNA)

Table 2.6 Approximate genomic size and average number of restriction enzyme sites in the genomes of different organisms. (1) A synthetic cloning vector. Synthetic DNA is often named with a p for plasmid, followed by identifying letters and the edition number. (2) Simian virus 40, a virus that infects monkeys. (3, 4) Bacteriophages are viruses that infect bacteria. (5) E. coli. (6) Sarcomyces cerevisea. (7) Drosophila melanogaster. (8) Mouse and human are about the same (in this respect). The total length of the mammalian chromosomal DNA is almost 2 m in life size; if it were printed in this type it would be well over 10000 km.

DNA LIGASE

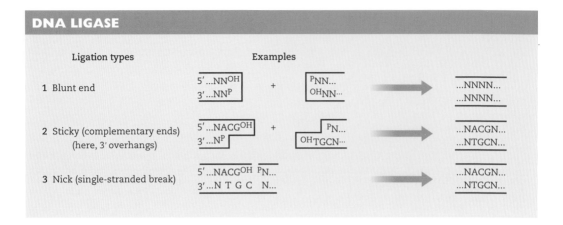

Ligation types   Examples

1 Blunt end

2 Sticky (complementary ends)
  (here, 3′ overhangs)

3 Nick (single-stranded break)

Fig. 2.10 DNA ligase catalyses the formation of a bond between juxtaposed 5′-phosphate and 3′-OH groups on the phosphate 'backbone' of DNA.

thousand base pairs, called CG dinucleotide 'islands', are found in which the CG dinucleotide occurs at nearly the expected frequency. These islands are often located 5′ of genes, an early observation that has since led to the identification and cloning of several genes based solely on their position 3′ of CG islands.

RESTRICTION AND MODIFICATION

Restriction enzymes are purified from a variety of bacteria, which use them as a simple defence against foreign DNA, particularly bacterio-phages. (Ironically, molecular biologists put their recombinant DNA plasmids into bacteria to 'grow-up' more.) Bacteria avoid having their restriction endonucleases cut their own genomic DNA by modifying it, particularly through methylation which blocks many restriction enzymes, rather than forbidding these DNA sequences. Bacteriophage DNA is said to be *restricted* (cut) by a bacterial strain when it cannot infect that strain of bacteria. This happens when the DNA has not been specifically *modified* and thereby protected against those restriction enzymes, although it may be able to infect another strain with different restriction endonucleases. Mammalian DNA is usually cut by restriction enzymes *in vitro* because just as bacterial DNA-modification systems vary amongst strains, eukaryotic methylation patterns differ from those of bacteria (prokaryotes). (Incidentally, methylation of DNA in eukaryotic cells is involved with compacting DNA and gene expression: it has nothing to do with protection from restriction enzymes.) In the laboratory, bacterial strains mutant in one or more of the *restriction–modification* (R–M) system enzymes are used as hosts to reduce incompatibility problems. All commonly used restriction enzymes are type II, which differ from the rarely used type I and III R–M systems in that they cut within the symmetrical DNA recognition sequence, are composed of a single polypeptide dimer (and therefore much easier to make than are multisubunit enzymes) and leave 3′-OH and 5′-phosphate ends.

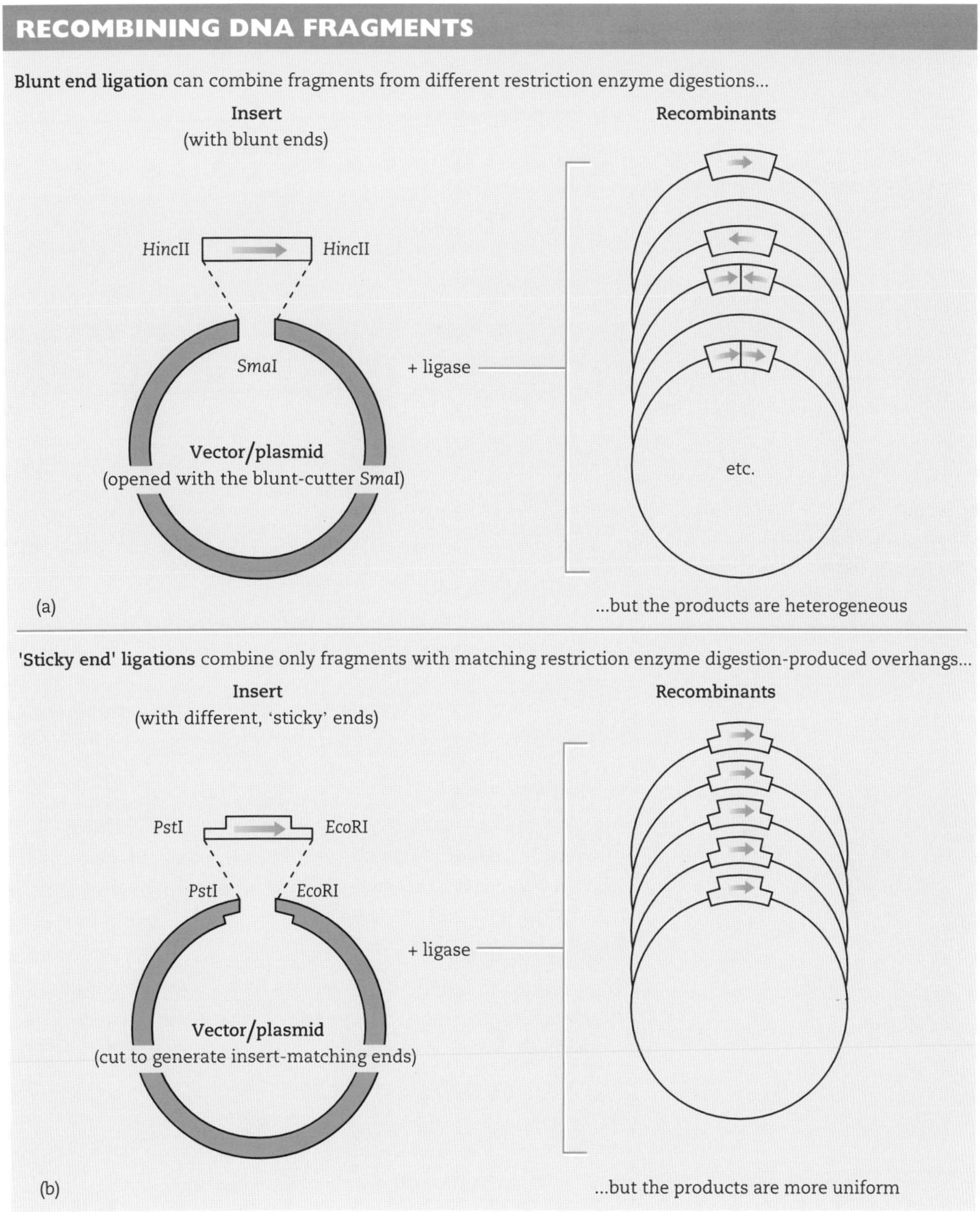

RECOMBINING DNA FRAGMENTS

Blunt end ligation can combine fragments from different restriction enzyme digestions...

Insert
(with blunt ends)

Recombinants

HincII → HincII

SmaI   + ligase

Vector/plasmid
(opened with the blunt-cutter SmaI)

etc.

(a)   ...but the products are heterogeneous

'Sticky end' ligations combine only fragments with matching restriction enzyme digestion-produced overhangs...

Insert
(with different, 'sticky' ends)

Recombinants

PstI → EcoRI

PstI   EcoRI

+ ligase

Vector/plasmid
(cut to generate insert-matching ends)

(b)   ...but the products are more uniform

Fig. 2.11 Recombining DNA fragments: mixing molecules and matching ends. Note that the blunt-cut vector (a) can close *without* an insert. This unwanted product of self ligation can be reduced by dephosphorylation (treatment with a phosphatase) of the vector. The sticky end vector (with different ends (b)) cannot close on itself because the overhanging ends are not compatible (they do not anneal to one another). Sticky end ligation with different ends permits 'directional cloning', since the insert can only 'go in' one way.

Ligases

Sticky ligations are more efficient because the compatible ends stick together (anneal), albeit weakly (Fig. 2.10). On the other hand, blunt end ligations are particularly useful because of the ability to recombine any DNA fragments, without a need for compatible ends (Fig. 2.11). DNA ligase also heals 'nicks' in DNA, where *one* of the two phosphate backbone chains is broken (Fig. 2.10).

EXTENDING AND CHEWING DNA (POLYMERASES AND NUCLEASES)

Polymerases

There are many different DNA polymerases (Pols), sometimes several per organism, but they have one activity in common (Fig. 2.12).

DNA Pol rules:

• all polymerases require an end to extend, none can synthesize DNA *de novo*;
• all polymerases extend the 3'-OH end, no DNA Pol adds to the 5' end;
• polymerases make a reverse copy (the *complement*) of the template strand.

The last rule is not absolute because there *are* enzymes that polymerize DNA and work without a template strand, but these are used relatively infrequently (e.g. terminal deoxytransferase (TdT), which adds a 'tail' of nucleotides). The first two rules of DNA Pols might make you think again about how genomic DNA is replicated. DNA Pol cannot start synthesizing DNA *de novo*, so instead replication of genomic DNA starts with RNA primers of ~10 bp length, which are first extended and then replaced with DNA by DNA Pol. Both strands are replicated as the 'replication fork' moves in one direction. Replication in the 5' → 3' direction is easy because that is the direction of DNA polymerization. In the 3' → 5' direction, small pieces of DNA (1000 to 2000 bp) are synthesized in the 5' → 3' direction and then ligated together.

ADDITION OF NUCLEOTIDES

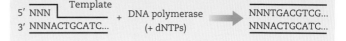

Fig. 2.12 *DNA Pols add to the 3' end of the DNA nucleotides that match the opposite strand (the template).*

Nucleases

Nucleases cut nucleic acids either into big pieces (*endo*nucleases) or many little pieces (*exo*nucleases). Exonucleases can be seen as undoing the work of polymerases, they depolymerize. Unlike the relatively uniform world of polymerases, however, nucleases demonstrate a wide range of abilities.

Nuclease anarchy

Nuclease anarchy.
• Some *exo*nucleases chew from one end (5′ → 3′).
• Some *exo*nucleases chew from the other end (3′ → 5′).
• Some *endo*nucleases take a bite right out of the middle, without starting at an end.
• Some nucleases strongly prefer single-stranded DNA or RNA. This preference forms the basis of the nuclease protection assay (see below).

Many DNA Pols also possess exonuclease activities. If present, a 3′ → 5′ exo activity (reverse of the 5′ → 3′ polymerase activity) can remove a newly added nucleotide. This activity is called 'proofreading' because it improves the fidelity of replication, perhaps because the polymerase is slower to extend a mismatched base (it is not base paired to the template, making it appear single stranded) and that gives the exo activity more time to digest. A 5′ → 3′ exo activity can clear the way ahead for the polymerase activity, digesting the old strand. This activity probably removes the RNA primer that initiates DNA replication.

DNA SYNTHESIS

Synthetic oligonucleotides (oligos) are made in the opposite direction. The chemical reaction proceeds by adding to the 5′ end, whereas the enzymatic reaction (polymerases) adds to the 3′ end.

DNA synthesis

Biology (polymerases)	5′-NNNN-3′ → extension
Chemistry (synthetic oligos)	extension ← 5′-NNNN-3′

This explains the desirability of gel-purifying synthetic oligos to ensure that they are full length before using them in certain procedures, especially those with single base resolution (sequencing), or even for PCR if the product will be cloned and the exact end is important (e.g. it recreates the restriction enzyme site). Even with a good (99%) yield for the addition of each nucleotide, the yield of full length oligo of 25 nucleotides (a '25mer') is only about 75% ($0.99^{25} = 0.778$), although each individual contaminant is present in very low amounts. Fortunately, the natural way to make DNA is much more efficient.

MAKING RNA FROM DNA AND VICE VERSA: RNA POL AND REVERSE TRANSCRIPTASE

RNA Pols make RNA from a DNA template (transcription).

RNA polymerase

There are three (mammalian) RNA Pols responsible for reading the DNA code and turning it into RNA, which either does the real work itself or is translated into protein (Fig. 2.13).

• RNA Pol I transcribes the rRNA genes.

• RNA Pol II transcribes the protein-coding genes, making mRNA.

• RNA Pol III transcribes the tRNA genes.

Pol II is the subject of the most intensive study because of its role in the regulation of protein expression. Transcription is a major point of regulation of gene expression and the rate of transcription is largely regulated by the binding of RNA Pol II to the gene promoter. Since RNA Pol II must transcribe many different genes and must therefore be relatively non-specific in

RNA POLYMERASE TRANSCRIPTION

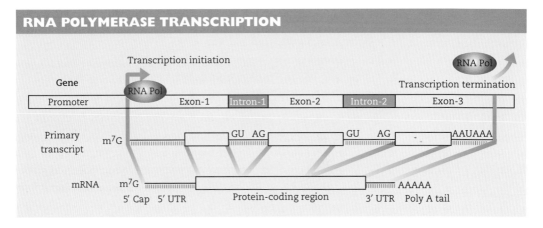

Fig. 2.13 RNA Pol transcription of an idealized gene. The protein-coding region is the part that is translated into protein. The protein-coding region is interrupted by introns, which are spliced out, and flanked by 5' and 3' untranslated regions (UTRs) that contain control elements regulating mRNA translation and stability. The promoter and other regions of the gene contain sequences that bind proteins that regulate transcription. The expression of a gene can also be controlled by the rate of splicing and alternative splicing. Splice donors have a GU pair at the start of the intron, and splice acceptors have an AG pair. The 7-methyl-guanosine (m^{7}G) cap is added to the mRNA almost immediately upon transcription initiation, whereas the poly (A) tail is added following the signal AAUAAA at the end of the mRNA sequence.

its favours, it is helped to identify exactly *which* gene should be transcribed and *when* by a host of gene-specific proteins that bind to sequences in the gene, called enhancers or silencers, that act to increase or decrease the frequency of transcription. Additional proteins help RNA Pol II to find exactly the correct start (initiation) and end (termination).

READING THE HELICAL DNA TEMPLATE

It is not known exactly how RNA Pol manages to read the helical DNA template, analogous to reading an inscription circling round a column, without getting the ever-lengthening RNA transcript hopelessly tangled. The current guess is that DNA helps by twirling around a nicked strand, thereby rotating beneath the RNA Pol. One reason this is thought to be true is because in bacteria the RNA message is already being translated before transcription is complete and it seems too improbable that the whole RNA Pol–mRNA–ribosome complex could twirl around the DNA. The rate of transcript extension by RNA Pol also becomes an important point when confronted with the size of some genes. For example, transcribing the 31.5 kbp mammalian dihydrofolate reductase (DHFR) gene at the rate of bacterial RNA Pol (60 bp/second) would take nearly 9 minutes.

Reverse transcriptase

RT makes DNA from an RNA template.

In the beginning, the word was that genetic information could go in only one direction, from DNA to RNA to protein. The dogma was shaken when it was found that viruses with a RNA genome (retroviruses) replicate through a DNA intermediate. The descriptive name *reverse transcription* was given to this activity, and the responsible enzymes are called RT (Fig. 2.14).

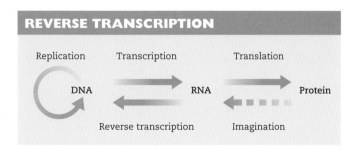

Fig. 2.14 Reverse transcription with the enzyme RT.

The ability to reverse transcribe mRNA into cDNA helps in gene cloning and recombinant protein production because the introns are eliminated, often greatly reducing the size of the DNA to be cloned and expressed. Also, recombinant protein encoded by cDNA can be produced in bacteria, which do not have introns and do not have the splicing mechanisms to remove introns. Using cDNA, one can clone and analyse the protein-coding part of a gene and then proceed directly to producing the protein in large quantities (see cDNA cloning and Recombinant proteins, Chapter 3).

CHAPTER 3

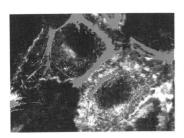

Molecular techniques

Techniques that are commonly used to analyse DNA, RNA and DNA-binding proteins are introduced in this chapter. General methods that are used for both DNA and RNA are discussed first, followed by particular methods that are employed in analyses directed towards answering specific questions.

BLOTTING

Blotting is a descriptive term for the *transfer of molecules out of a gel and onto a filter membrane by wicking action* (Fig. 3.1), although the term is now used for electro- or vacuum transfers, and even for simply binding molecules to filters that were not gel resolved ('dot blots'). Blotting was developed to make the gel-resolved nucleic acid more accessible to subsequent manipulation, such as identification by hybridization (see next section).

In a Southern blot, DNA is separated on a non-denaturing gel, denatured with sodium hydroxide and blotted. The blot is then probed with either labelled DNA or RNA. The gel may also be a denaturing urea/polyacrylamide gel, in which case

BLOTTING

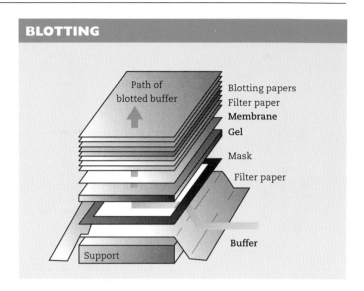

Fig. 3.1 Blotting nucleic acids onto membranes. The support keeps the stack of filter papers, gel, membrane and blotting paper out of the buffer bath. The blotting papers absorb buffer. Wicking action (absorption) pulls the buffer up through the gel, carrying the nucleic acids along, and then through the membrane, where the nucleic acid binds. The mask ensures that the buffer goes through the gel rather than taking the easier route around it.

denaturation prior to transfer is not necessary. The Southern blot is named after its originator, Dr Edwin Southern.

In a Northern blot, RNA is separated on a denaturing agarose gel (containing formamide/formaldehyde) and blotted. The blot is then probed with labelled DNA or RNA. The Northern blot was named *after* the Southern blot.

In a Western blot, proteins are separated on a denaturing sodium dodecyl sulphate (SDS)/polyacrylamide gel, and blotted then probed with antibodies. To complete the naming, DNA-binding proteins are detected in a Southwestern blot and RNA-binding proteins are detected in a Northwestern blot. In these assays, proteins are separated and then probed with nucleic acids. The future directions in blotting techniques seem clear.

HYBRIDIZATION

Hybridization is the formation of base pairs between nucleic acids. The term was originally restricted to the formation of *hybrid* double-stranded molecules, as in a Northern blot where DNA/RNA hybrid molecules are formed. Typically, a mixture of nucleic acids is resolved on a gel, blotted onto a nitrocellu-lose or nylon membrane, and the blot is hybridized with a labelled (easily detectable) piece of DNA or RNA (the probe, Fig. 3.2). The probe sticks to its complementary strand, the blot is washed to remove unbound and non-specifically bound

NUCLEIC ACID BLOTTING

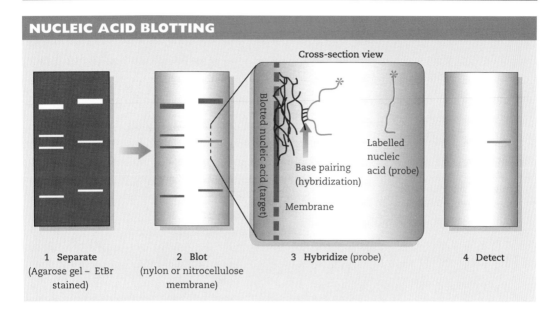

Cross-section view

1 Separate
(Agarose gel – EtBr
stained)

2 Blot
(nylon or nitrocellulose
membrane)

3 Hybridize (probe)

4 Detect

Blotted nucleic acid (target)

Base pairing
(hybridization)

Labelled
nucleic
acid (probe)

Membrane

Fig. 3.2 Nucleic acid blotting. The method of detection depends on the type of label on the probe. Typically, the probe is radioactively labelled and detected on photographic film (autoradiography).

probe and the remaining probe is then detected. The method of detection depends on how the probe was labelled. Probes are often labelled with radioactive nucleotides and the probe is detected by autoradiography (self-exposure on film).

The longer the length of hybridized strands, the stronger the binding (up to about 100 bp). By using more or less stringent conditions to wash away unbound and weakly bound probe, it is possible to determine how well the probe is bound to the blotted target nucleic acid (Table 3.1). High temperature and formamide tend to make the hybridized strands come apart, whereas high salt concentration stabilizes the hybrid by shielding the negative phosphate charges.

So, a high-stringency wash is high temperature, high formamide and low salt concentration. *A blot is hybridized at high stringency to detect only the nucleic acids that are very closely related or identical to the probe. A blot is probed (hybrid-*

Table 3.1 The hybridizing washing conditions for high and low stringency blotting

HIGH AND LOW STRINGENCY BLOTS

Hybridize/wash conditions	Low stringency	High stringency
Temperature	Low	High
Formamide concentration	Low	High
Salt concentration	High	Low

ized and washed) at *low* stringency to detect more distantly related sequences. This might be done to identify additional members of a multigene family or homologous genes in other species.

Oligonucleotides (oligos) may be used in detecting blotted nucleotides. Base pairing with the complementary nucleic acid is also called *annealing*, a term that is used especially often for synthetic oligonucleotides. Hybridization protocols must be specifically adapted to oligos because they are usually shorter than probes derived from DNA fragments and, therefore, the hybrids have lower stability and the sensitivity is reduced. The advantages of oligoprobes are that they are easy to make (just fax the sequence to your favourite biotech company), easy to label and strand-specific (e.g. antisense sequence would be used to detect messenger RNA (mRNA, p. 76)).

TRANSFECTION AND TRANSFORMATION: SELECTION AND SCREENING

Transfection and transformation are ways of putting genes into cells. Selection and screening are ways of finding the cells that received the genes.

Two different names are given to the process of putting genes back into cells. Bacterial cells are said to be *transformed* because in a classic experiment non-pathogenic bacteria were made pathogenic by addition of the 'transforming principle', i.e. DNA. Mammalian cells are *transfected* because the transfer of genes is similar to viral infection. Both cell types spontaneously pick up DNA, but the efficiency is very low. Numerous transfection techniques have been developed to increase the efficiency of uptake, such as the following.

• Calcium phosphate precipitation—forms aggregates of DNA that precipitate out of solution and are endocytosed by the cells.

• DEAE-dextran—an anion-binding gel that forms large aggregates with DNA.

• Cationic liposomes—coats the (anionic) DNA in lipid so that it passes more readily through the cell membrane.

• 'Biolistics'—delivery into the cell of DNA coated onto gold microprojectiles.

• Electroporation—a high-voltage shock that makes transient, DNA-permeable holes in the cell membrane.

In spite of these advances, a good transfection of mammalian cells results in only ~10% of the cells receiving DNA and < 1% of these are stable transfectants, with the DNA incorporated into the chromosomes. Many experiments are designed to find out what you want to know *before* the DNA is lost from the cell, in the time between the 10% initial transfection rate and < 1% stable transfection. These are called *transient transfections* (or simply transients) because they last only a few days. Since transient transfections do not involve long-term culture, they are especially useful when many different DNA constructs are to be tested, such as when regulatory elements of a gene are being located, or when the recipient cell is not able to be cultured for a long period, as with a primary culture from an untransformed tissue.

With the low efficiency of transformation and transfection, the second problem is to identify the cells that have taken up DNA. This is accomplished by either *screening* or *selecting* positive clones, or by a combination. An example of selection is when bacteria are transformed with a plasmid containing a gene that confers antibiotic resistance. The few cells that take up the plasmid and express the gene will be able to grow in medium containing the antibiotic, e.g. ampicillin, whilst the other cells will not (Fig. 3.3). Similarly, mammalian cells stably transfected with a gene for resistance to a toxin also can be selected in a toxin-containing medium. G418 or hygromycin B are examples of these toxins.

Alternatively, the expression of a plasmid-encoded enzyme in the cell might be easy to detect even though it does not change the growth characteristics or confer any resistance. This is *screening* for expression. A commonly used enzyme is β-galactosidase (β-gal), whose activity can be assayed with chromogenic substrates that either remain soluble or precipi-

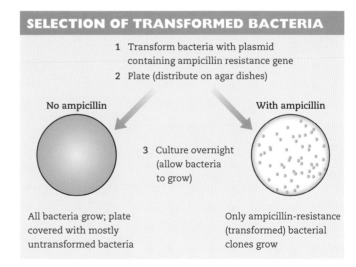

SELECTION OF TRANSFORMED BACTERIA

1 Transform bacteria with plasmid containing ampicillin resistance gene

2 Plate (distribute on agar dishes)

No ampicillin

With ampicillin

3 Culture overnight (allow bacteria to grow)

All bacteria grow; plate covered with mostly untransformed bacteria

Only ampicillin-resistance (transformed) bacterial clones grow

Fig. 3.3 Selection of transformed bacteria by their acquisition of the ability to grow on agar containing an antibiotic (antibiotic resistance). The transforming plasmid encodes the antibiotic resistance gene, in this case resistance to ampicillin.

tate. For example, the β-gal-catalysed conversion of a colourless compound called Xgal (the real name is very long), to a blue precipitate allows the identification of β-gal⁺ bacteria or animal cells (Fig. 3.4).

SCREENING FOR RECOMBINANT DNA PLASMIDS

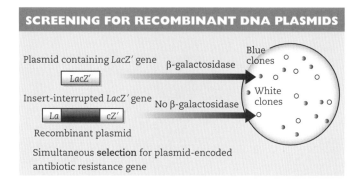

Fig. 3.4 Screening for β-gal expression after selecting for antibiotic resistance. Bacteria are cultured on agar plates in the presence of Xgal (chromogenic substrate of β-gal) and IPTG (inducer of β-gal expression).

Many experiments call for a combination of selection and screening. The plasmids used in cloning genes usually confer resistance to antibiotics (selection) and contain a β-gal gene that is disrupted by the recombinant genes (screening, Fig. 3.4). The clones that contain recombinant DNA have an interrupted *lacZ'* gene, do not make β-gal and do not turn blue when plated on Xgal. Only these 'white' clones are picked and further tested to determine whether the insert is correct. Note that these subsequent tests are often also a form of screening, e.g. looking for particular restriction enzyme sites, but they are much more difficult and slower. A powerful early screen saves time, just how much time is saved depends on the frequency of 'positive' clones.

Another common combination of selection and screening occurs when making monoclonal antibodies in mammalian cells (p. 116). Hybridoma cell lines are cell fusions (hybrids) that secrete monoclonal antibodies. They are generated when an antibody secreting normal (mortal) cell is fused *in vitro* with an immortalized cell line that is mutant in the *hypoxanthine guanine phosphoribosyl transferase* (HGPRT) gene. (This gene is encoded on the X chromosome, so there is only one active copy per cell.) The hybrids receive a functional HGPRT gene from the normal cell and can thus grow in a medium that selects for HGPRT function, whilst the otherwise-immortal cells are killed in this medium. Only after the hybridoma cell lines are selected for growth are they then screened for antibody secretion.

POLYMERASE CHAIN REACTION (PCR)

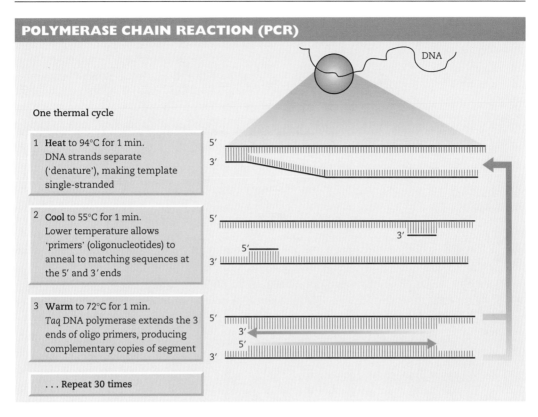

One thermal cycle

1 **Heat** to 94°C for 1 min.
DNA strands separate
('denature'), making template
single-stranded

2 **Cool** to 55°C for 1 min.
Lower temperature allows
'primers' (oligonucleotides) to
anneal to matching sequences at
the 5' and 3' ends

3 **Warm** to 72°C for 1 min.
Taq DNA polymerase extends the 3
ends of oligo primers, producing
complementary copies of segment

... Repeat 30 times

Fig. 3.5 Some like it hot. Steps 1–3 double the amount of template.
Repeated 30 times (more or less), they yield a million-fold
amplification in about 1 hour.

DNA ANALYSIS

Polymerase chain reaction (PCR)

PCR is used to amplify
fragments of DNA.

This simple but incredibly powerful technique has revolution-
ized molecular biology and many applications have already
found their way into clinical and forensic medicine. One can, in
principle, generate any amount of DNA from a single molecule.
The procedure is incredibly simple, just add the ingredients
(template, oligoprimers, nucleotides, buffer, polymerase) and
bake (and cool, bake again, and cool . . . , Fig. 3.5).

PCR amplifies the segment of DNA between the
oligonucleotide primers, which are usually 15–25 bp long and
designed to match specific sequences (5' and 3') flanking the
segment to be amplified. Therefore, it is *usually* necessary to
know these flanking sequences. (Methods have been developed to allow PCR amplification of segments when only one

end is known: anchored PCR where a 'tail' of cytidine 5′-triphosphate (dCTP) is added to the unknown end, or ligation mediated PCR where a double-stranded oligonucleotide is ligated to the unknown end. In both methods a known sequence is placed at the unknown end of the fragment to be amplified.)

If the amplification were 100% efficient, then the number of template molecules would double every cycle ('exponential' phase amplification) and 30 cycles would produce over a thousand-million-fold amplification (2^{30} = 1 073 741 824, Fig. 3.6). Unfortunately, even PCR is not perfect. The latter cycles often produce linear amplification ('plateau' phase, see below) as the reagents are used up, or begin to suffer from being cooked and cooled so often or simply from competition for the polymerase molecules (Fig. 3.7). Even with considerably lower overall efficiency, however, one can easily generate enough product to see on an ethidium bromide (EtBr)-stained agarose gel starting with a single copy target sequence from 1 ng of genomic DNA. This is the DNA content of approximately 2000 (nucleated) mammalian cells, which can be found in <0.1 ml of blood.

The original PCR was performed with a 'regular' DNA polymerase (Pol). (It was actually a piece of DNA Pol, the so-called Klenow fragment, which is missing the exonuclease activity of parent enzyme.) Fresh enzyme had to be added for each thermal cycle because the enzyme was irreversibly denatured at the temperatures required to melt the DNA. This flaw in an otherwise beautiful procedure spurred the characterization and large-scale production of *heat-stable enzymes*,

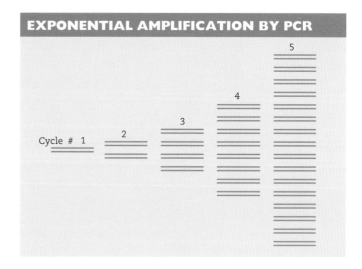

EXPONENTIAL AMPLIFICATION BY PCR

Cycle #   1   2   3   4   5

Fig. 3.6 Exponential amplification of a DNA template by PCR. Only five perfectly efficient cycles are shown. At this scale, representing a million-fold amplification would require approximately 3 km.

PRODUCT ACCUMULATION DURING PCR

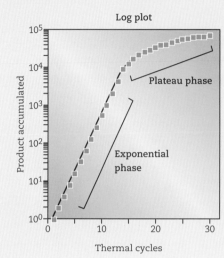

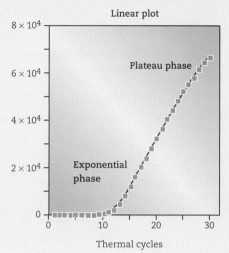

Fig. 3.7 Product accumulation during PCR as a function of cycle number plotted either on logarithmic scale (left) or linear scale (right). PCR is often biphasic: exponential amplification in the early cycles followed by non-exponential amplification in the later cycles, called the plateau phase. The point at which the reaction becomes non-exponential depends on the reaction conditions. Note that the significant accumulation of product during the later cycles seems to belie the name 'plateau phase'.

the first of which to be widely used was *Taq* DNA Pol (from *Thermus aquaticus*). Since then, several additional DNA Pol, as well as RNA Pol and DNA ligases have been offered commercially.

PCR VARIATIONS

PCR has changed fundamentally the way in which molecular biology is done. Seemingly infinite variations of this basic reaction are still being developed. One of the most useful and generally applicable variations is called 'hot start', in which a critical reaction component is withheld until the reaction is hot (⩾60°C), when the component is added and the reaction starts. This means that the stringency of the primer annealing is high when the DNA Pol is allowed to extend the primer, greatly increasing the specificity. This protocol also illustrates the distinction between specificity of detection, which is optimal at high temperatures, and amplification of the sequence, which may be more efficient at lower temperatures. The best of both worlds is found in protocols that use high oligoannealing temperatures at the beginning of cycling to enforce specificity and then gradually lowering the annealing temperature to encourage amplification.

Future amplification systems may be non-thermal cycling (isothermal). Room temperature or 'warm' amplification/ detection systems may have particular application in clinical testing, where time is critical and thoroughly tested reagent kits can be prepared for specific assays. For example, the amplification of pathogen-specific DNA or RNA may allow the

ISOTHERMAL DETECTION/AMPLIFICATION SYSTEMS

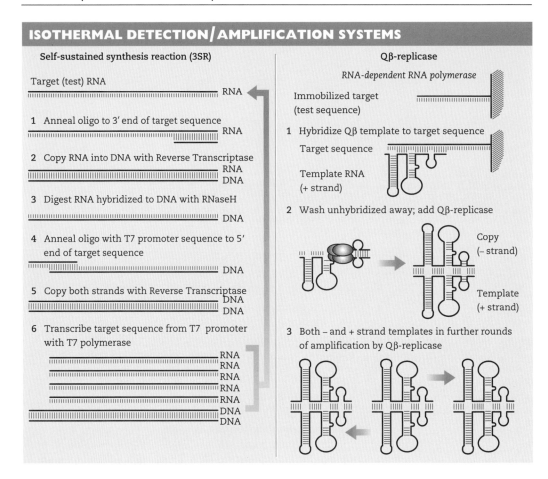

Self-sustained synthesis reaction (3SR)

Target (test) RNA

—————————————————————————————— RNA

1 Anneal oligo to 3′ end of target sequence

—————————————————————————————— RNA

2 Copy RNA into DNA with Reverse Transcriptase

—————————————————————————————— RNA
—————————————————————————————— DNA

3 Digest RNA hybridized to DNA with RNaseH

—————————————————————————————— DNA

4 Anneal oligo with T7 promoter sequence to 5′
   end of target sequence

—————————————————————————————— DNA

5 Copy both strands with Reverse Transcriptase
—————————————————————————————— DNA
—————————————————————————————— DNA

6 Transcribe target sequence from T7 promoter
   with T7 polymerase

—————————————————————————————— RNA
—————————————————————————————— RNA
—————————————————————————————— RNA
—————————————————————————————— RNA
—————————————————————————————— RNA
—————————————————————————————— DNA
—————————————————————————————— DNA

Qβ-replicase

RNA-dependent RNA polymerase

Immobilized target
(test sequence)

1 Hybridize Qβ template to target sequence

Target sequence

Template RNA
(+ strand)

2 Wash unhybridized away; add Qβ-replicase

Copy
(– strand)

Template
(+ strand)

3 Both – and + strand templates in further rounds
   of amplification by Qβ-replicase

Fig. 3.8 Some like it cooler: two isothermal detection/amplification systems. (left) All of the reactions occur simultaneously on different molecules during the 3SR. Amplification specificity is provided by the sequence of the oligonucleotides that are annealed in steps 1 and 4. Only the RNA that is hybridized to DNA is digested by RNase H (hybrid) in step 3. (right) Different target sequences can be hybridized to the Qβ-replicase template to test for the presence of particular RNAs. Good templates for Qβ-replicase possess large amounts of secondary structure which probably destabilize the post-replication (–/+) H, allowing the strands to act as templates in further rounds of amplification.

rapid identification of infections or monitoring of a patient's response to treatment. Two especially interesting amplification methods are based on coupled RNA Pol–reverse transcriptase (RT) reactions (the self-sustained synthesis reaction, 3SR) and Qβ-replicase (Fig. 3.8).

The 3SR enzyme system is modelled on retrovirus replication and can produce a 10-million-fold amplification in <2 hours. Qβ-replicase is a RNA Pol derived from a bacterio-

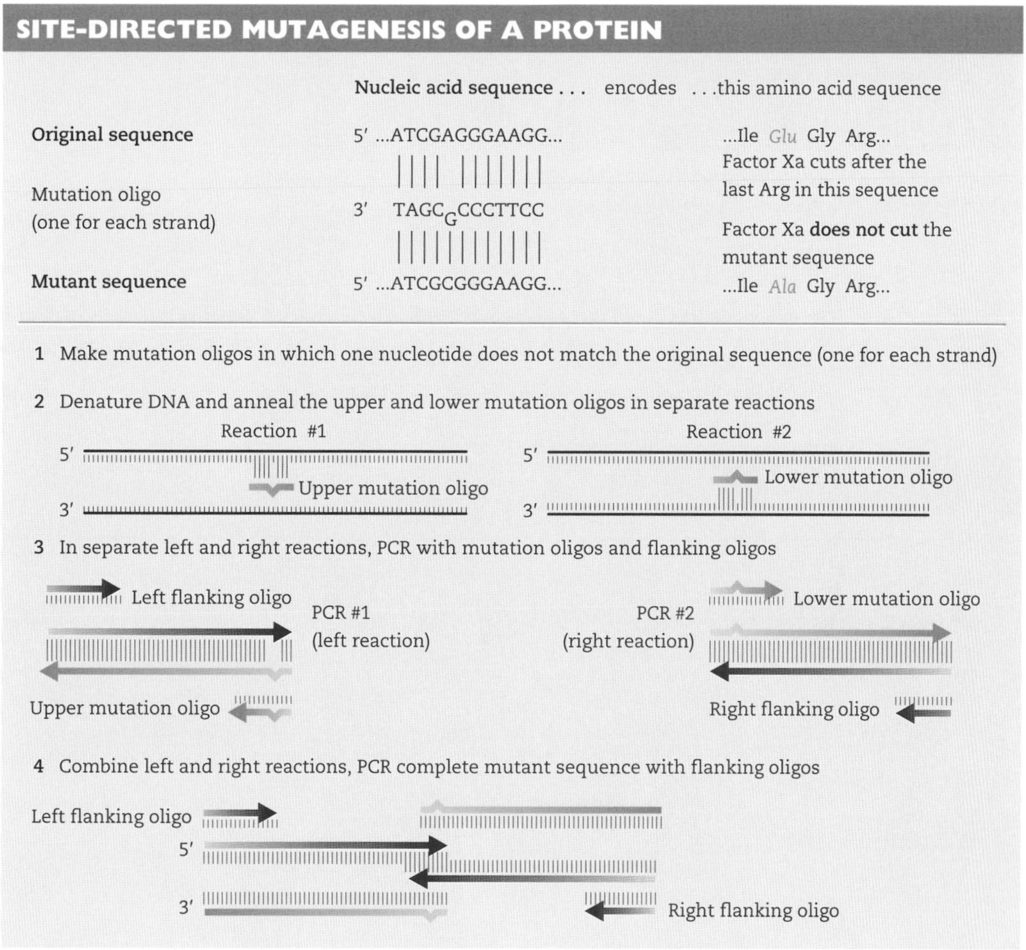

SITE-DIRECTED MUTAGENESIS OF A PROTEIN

Nucleic acid sequence . . . encodes . . .this amino acid sequence

Original sequence	5′ ...ATCGAGGGAAGG...
	‖‖‖‖ ‖‖‖‖‖‖‖
Mutation oligo (one for each strand)	3′ TAGC$_G$CCCTTCC
	‖‖‖‖‖‖‖‖‖‖‖‖
Mutant sequence	5′ ...ATCGCGGGAAGG...

...Ile *Glu* Gly Arg...
Factor Xa cuts after the
last Arg in this sequence

Factor Xa **does not cut** the
mutant sequence
...Ile *Ala* Gly Arg...

1 Make mutation oligos in which one nucleotide does not match the original sequence (one for each strand)

2 Denature DNA and anneal the upper and lower mutation oligos in separate reactions

3 In separate left and right reactions, PCR with mutation oligos and flanking oligos

4 Combine left and right reactions, PCR complete mutant sequence with flanking oligos

Fig. 3.9 Site-directed mutagenesis of a protein. In this example, the recognition site for the complement enzyme, factor Xa, is mutated. The resulting protein is not cleaved and might, therefore, prove to be more stable *in vivo*.

Site-directed mutagenesis is used to prepare specific mutants (variants) of a known gene.

phage that can produce a 100-million-fold amplification in 15 minutes. Refined, such methods might make the coffee-maker the only thermal cycler in the laboratory or clinic.

Site-directed mutagenesis

At last, here is an art, in the form of biotechnology, that improves on nature. Natural mutants are generated *randomly*, through mistakes in DNA synthesis or repair, and most are deleterious. The rare beneficial mutation could still be lost unless it happens to be really useful, such as when the alternative is extinction. Even with accelerated mutagenesis and screening/selection *in vitro*, it makes for very slow going, far too slow for your biotechnologist-on-the-go. Synthetic oligonucleotides are the answer for quick (and with PCR, easy) designer genes (Fig. 3.9).

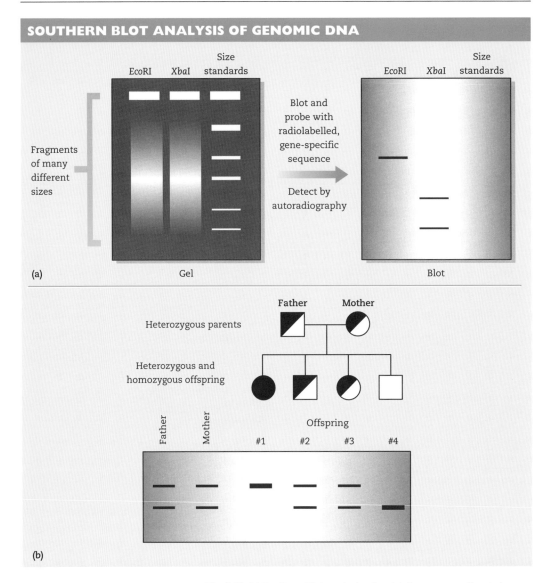

Fig. 3.10 (a) Southern blot analysis of restriction enzyme digested genomic DNA. On the left is the EtBr-stained agarose gel showing the heterogeneously sized fragments of restriction enzyme digested genomic DNA (the average length is a function of the recognition site size). The gene is found on one *Eco*RI fragment and two smaller *Xba*I fragments, i.e. there is an *Xba*I site within the gene. (b) Southern blot analysis of the offspring between heterozygous parents. In this example, digestion of genomic DNA with a restriction enzyme produces different sized fragments. This is known as *restriction fragment length polymorphism* (RFLP). If the larger fragment (indicated with black in the pedigree) were associated with a disease, such an analysis would strongly suggest close monitoring of the first offspring (the daughter symbolized by a filled circle), that the second and third offspring are carriers, and that the fourth offspring does not carry the the disease-associated allele (the son symbolized by an open square).

Southern blotting

A Southern blot is performed to identify and to determine the size of a DNA fragment.

Southern analysis involves separating DNA fragments on a gel, blotting the gel and then hybridizing the blot with a labelled probe (Fig. 3.10a). The gel separates the DNA fragments according to size and the hybridization step provides specificity in detection, allowing the identification of DNA fragments. For example, to determine the size of the genomic EcoRI fragment(s) containing the β-globin gene, one would cut genomic DNA with EcoRI, separate the resulting fragments on an agarose gel, blot and probe with a fragment of the β-globin gene. Alternatively, to test whether a PCR-amplified product is what you think it is, you might blot it and probe the blot with a labelled fragment of the gene.

An important application of Southern blotting is in pedigree analysis (Fig. 3.10b). Testing whether offspring are carriers of genes that predispose to or cause disease, and paternity testing are two examples of this analysis. Genomic DNA is cut with a restriction enzyme, blotted and the blot is hybridized. Either the restriction enzyme or the probe distinguishes between the alleles of the gene.

DNA sequencing

DNA is sequenced to determine the order of the nucleotides.

There are two ways to sequence DNA, the hard way using noxious chemicals and the easy, modern way using enzymes and inexpensive kits. Both methods involve four base-specific reactions that remain incomplete, yielding a heterogeneous population of molecules that end at one of the four nucleotides in each of four separate reactions. The enzymatic method of sequencing uses dideoxynucleotides (ddNTPs), which differ from normal deoxynucleotides (dNTPs) in that they cannot be extended by DNA Pol. These populations of molecules are then resolved on a denaturing polyacrylamide gel, one lane per base-specific reaction (Fig. 3.11).

A modification of the DNA sequencing technique uses thermal-stable DNA Pol, as used in PCR, and repeatedly anneals, extends and melts. Although no chain reaction is established in this procedure because the product does not act as template in the next round, it can provide a sequence from much less DNA template. Importantly, the elevated temperature reduces interference from secondary structures in the template and makes the sequencing of double-stranded DNA easier.

DNA SEQUENCING

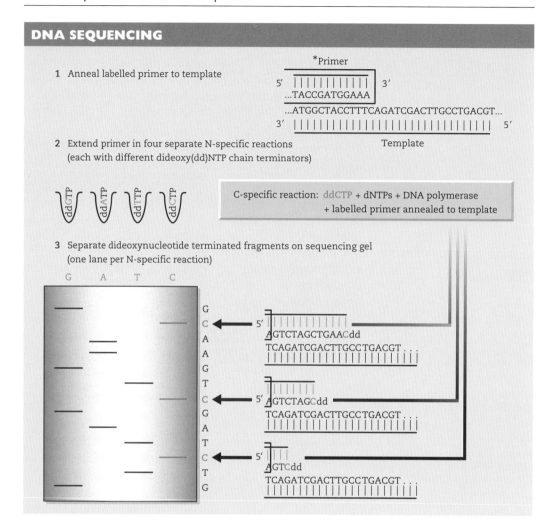

1 Anneal labelled primer to template

2 Extend primer in four separate N-specific reactions
   (each with different dideoxy(dd)NTP chain terminators)

C-specific reaction: ddCTP + dNTPs + DNA polymerase
                      + labelled primer annealed to template

3 Separate dideoxynucleotide terminated fragments on sequencing gel
   (one lane per N-specific reaction)

Fig. 3.11 DNA sequencing using dideoxy chain terminators (a.k.a. Sanger sequencing). Note that at each step, only *some* reactions are terminated by the addition of ddNTP. Otherwise, no sequence would be obtained beyond the first nucleotide.

In chemical sequencing, also called *Maxam–Gilbert* sequencing after the developers, sequence is obtained from the DNA itself, instead of an enzymatic copy. The DNA to be sequenced is labelled at one end and then treated with chemicals that specifically destroy one or two of the nucleotides (both purines or both pyrimidines or only one nucleotide). Again, these reactions are not allowed to go to completion, they remain incomplete reactions, or else no sequence 'ladder' would be generated.

RNA ANALYSIS

Northern blot

Northern blots are used to quantify the amount of mRNA and to determine mRNA size.

In a Northern analysis, RNA molecules of different sizes are

NORTHERN BLOT ANALYSIS

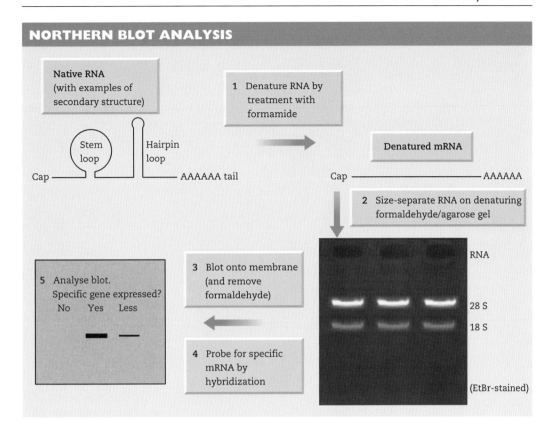

Fig. 3.12 Northern blot analysis of gene expression. Along with quantifying the *amount* of mRNA, mRNA *size* can be *estimated* by comparison to the ribosomal RNA (rRNA) bands, although size standards are customarily employed for size determination. The expression of similar genes (homologues) could also be tested by using less stringent probe hybridization conditions.

resolved on a denaturing formaldehyde-containing agarose gel, the gel is blotted and the blot is hybridized with a labelled probe (Fig. 3.12). The denaturing gel unfolds the RNA so that separation is based on length alone and not on secondary structure (e.g. hairpin loops). The formaldehyde in the gel reacts with the denatured nucleotides and prevents their refolding. The formaldehyde is soaked out of the gel and away from the RNA before blotting.

Nuclease protection assay

Nuclease protection assays are used to quantify mRNA and probe mRNA structure.

Certain nucleases can digest only single-stranded nucleic acids; double-stranded nucleotides are not digested. This forms the basis of the nuclease protection assay where mRNA, which is

NUCLEASE PROTECTION ASSAY

1 **Mix** labelled probe with RNA. Probe matches a segment of the RNA (label is indicated with ∗). RNA is a mixture of many sequences

2 **Hybridize** probe to RNA. Probe hybridizes specifically to matching (complementary) RNA sequence

Hybridized probe/RNA   Unhybridized probe

3 **Digest** unhybridized RNA and probe with nuclease. Hybridized probe and RNA are protected against nuclease digestion

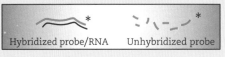

Hybridized probe/RNA   Unhybridized probe

4 **Analyse** protected fragments on sequencing gel

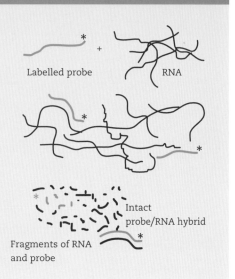

Labelled probe   RNA

Intact probe/RNA hybrid

Fragments of RNA and probe

Fig. 3.13 The nuclease protection assay provides a sensitive quantitation of mRNA. Excess probe is added and hybridized to the RNA (steps 1 and 2). The amount of the probe protected from digestion by nuclease depends on the amount of matching RNA (step 3). Intron/exon junctions and transcription start sites could also be mapped with probes that overlap the junctions.

single stranded, is hybridized to a single-stranded probe. The hybridized part of the probe is resistant to treatment with single-strand-specific nucleases (Fig. 3.13). There are two major types of nuclease protection, depending upon whether the probe is RNA or DNA.

1 In a *RNase protection* assay, the unhybridized part of a *RNA probe* is digested with RNase A and RNase T1.

2 In a *S1 nuclease protection* assay, the unhybridized part of a *DNA probe* is digested with S1 nuclease.

Although the nuclease protection assay entails more work than a Northern blot, it has certain advantages for analysis of mRNA levels, including:

• considerably *less RNA* can be used <2–5 μg versus 5–20 μg total RNA);

• it is *more quantitative*, because hybridization (but not blotting) is well defined in physical–chemical terms (kinetics, temperature of hybridization);

• mRNA fine *structure can be analysed* because the probes can be designed to distinguish even single nucleotide differences between target sequences;

• somewhat *degraded RNA can be analysed* because only a relatively small portion of it is hybridized (this is not an endorsement of sloppy RNA preparation).

cDNA CLONING

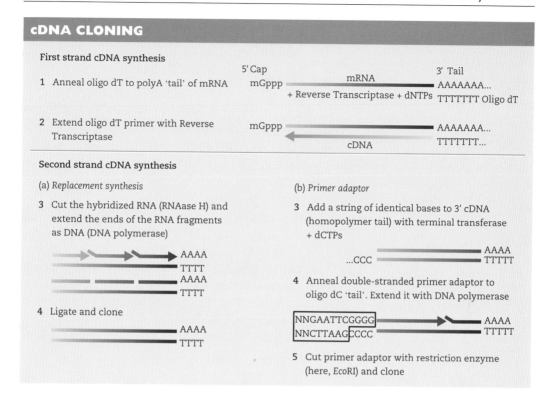

First strand cDNA synthesis

1. Anneal oligo dT to polyA 'tail' of mRNA

5' Cap   3' Tail
mGppp ————— mRNA ————— AAAAAAA...
+ Reverse Transcriptase + dNTPs TTTTTTT Oligo dT

2. Extend oligo dT primer with Reverse Transcriptase

mGppp ————— ————— AAAAAAA...
cDNA   TTTTTTT...

Second strand cDNA synthesis

(a) *Replacement synthesis*

3. Cut the hybridized RNA (RNAase H) and extend the ends of the RNA fragments as DNA (DNA polymerase)

————→————→————→ AAAA
————————————— TTTT
————————————— AAAA
————————————— TTTT

4. Ligate and clone

————————————— AAAA
————————————— TTTT

(b) *Primer adaptor*

3. Add a string of identical bases to 3' cDNA (homopolymer tail) with terminal transferase + dCTPs

————————————— AAAA
...CCC ————————— TTTTT

4. Anneal double-stranded primer adaptor to oligo dC 'tail'. Extend it with DNA polymerase

NNGAATTCGGGG ————→ ————— AAAA
NNCTTAAGCCCC ————————— TTTTT

5. Cut primer adaptor with restriction enzyme (here, EcoRI) and clone

Fig. 3.14 cDNA cloning. The first cDNA strand is usually primed by oligo-dT, which anneals to the poly(A) tail of mRNA. Second strand synthesis is more difficult and two of the many different protocols are shown. (a) Replacement synthesis is relatively simple and effective but it has the disadvantage that the extreme 5' end is lost because there is no primer further upstream. (It is not known how the mRNA cap is removed or why subsequent ligation into cloning vectors proceeds efficiently. Nevertheless, it works.) (b) The primer–adapter method can succeed in cloning the 5' end but it requires more steps and the terminal transferase step is hard to control (step 3). Note that for this method, the oligo-dT primer in the first strand synthesis often also contains a restriction site, which permits directional cloning into expression vectors (see Recombinant proteins).

Complementary DNA (cDNA) cloning

mRNA is the RNA that is translated by the ribosome into protein. cDNA is the reverse complement of mRNA. All mRNAs possess a poly(A) tail (a string of A ribonucleotides that is added to the 3' end of mRNA after transcription), whereas ribosomal RNA (rRNA) or transfer RNA (tRNA), which make up the bulk of total or cytoplasmic RNA preparations, do not. Oligo-dT (a string of T *deoxyribonucleotides*) is often used to prime the first strand synthesis by RT, because it anneals specifically to the poly(A) tail and can thereby selectively prime cDNA synthesis from mRNA (Fig. 3.14). The second strand is more difficult to synthesize and several methods have been developed.

Here are several reasons why cDNA cloning is popular and powerful.

• cDNA is *shorter* than the gene. Some relatively small proteins are encoded by incredibly large genes, possessing long introns and untranslated regions (UTRs). Dihydrofolate reductase (DHFR), for example, has a protein-coding region of <600 bp spread over a gene that is 31 500 bp.

A clone is a population of genetically identical organisms, cells or DNA molecules, derived from the replication of a single progenitor

DNA cloning is the production of multiple identical copies of a DNA fragment

• cDNA contains all the *important* parts. The exons are usually the more interesting part, at least initially, because their sequence affords what is often the first glimpse into what the protein looks like.
• cDNA can be made from RNA that is *enriched* for the mRNA of interest by using particularly high-expressing cells or by treating cells to induce the mRNA, whereas for the genomic sequence only two copies are present in each and every (somatic) cell.

Recombinant proteins

Recombinant DNA technology allows the production of large quantities of pure proteins for clinical practice and research (p. 110). Note that although it is the DNA that is recombinant, the proteins produced by these techniques are often termed recombinant even when they are identical to the naturally produced proteins. The recombinant human proteins used in clinical practice include the following.
• Insulin (the first recombinant protein used in clinical practice).
• Growth hormone.
• Erythropoietin. (EPO, a hormone produced in the kidney which stimulates red cell production. Patients with renal failure are deficient in EPO.)
• Factor VIII, a clotting factor. (Haemophiliacs are deficient in factor VIII.)
• Tissue-type plasminogen activator (t-PA, a thrombolytic enzyme).

Once the gene for a particular protein has been cloned it is inserted into plasmid under the regulation of a strong promoter. Such a construct is called an *expression vector* because it is designed to express large amounts of protein. Different promoter sequences are used depending upon whether the protein is to be made in bacterial, yeast or mammalian cells. The expression vector encoding the recombinant protein is then stably transfected into the genome of microorganisms or cultured mammalian cells, which then produce the protein.

RT–PCR assay

RT–PCR is used to detect very small amounts of mRNA.

Retroviruses use reverse transcriptase (RT) to convert their RNA genomes into DNA. So, what can RT do for you? When coupled with PCR amplification, it can provide the only means of measuring very small amounts of RNA, much less than is necessary for a Northern blot or even a nuclease protection assay. The RNA is reverse transcribed by RT using six-base oligonucleotides with random sequence ('random oligos') that anneal all over the RNA and act as primers to produce a cDNA strand, which is then amplified by PCR.

RT–PCR is excellent for determining whether a gene is transcribed and it can be used to measure large differences in mRNA levels, but it is poor at reliably measuring smaller differences (10-fold and less). This is because slight differences in the efficiency of PCR amplification produces large differences in product, independent of the original amount of template. For example, the difference between an efficiency of 80 and 85% yields a difference of more than sixfold after 30 cycles $(1.6^{30} = 1.3 \times 10^6$ versus. $1.7^{30} = 8.2 \times 10^6)$. A modification, called *competitive PCR*, controls for differences in amplification efficiency by including a titred amount of a distinguishable template in every sample (Fig. 3.15). The point in the titration at which the cDNA and the added competitor template produce an equal amplification is then taken as a measure of the original cDNA amount.

Note that the absolute amounts of the competitor and cDNA are not necessarily the same at the equivalence point, they are only amplified to yield equivalent amounts of product. Different efficiencies of amplification may occur when the sequences are different (one is preferentially amplified), perhaps because it contains lower G/C content (and is easier to

1 *Make cDNA* from RNA with RT and *random oligoprimers.*
2 *PCR amplify* template cDNA using *gene-specific* oligoprimers.
3 *Gel analyse* PCR products.

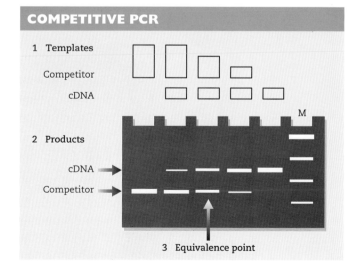

Fig. 3.15 Competitive PCR. (1) PCR with a constant amount of cDNA and a titred competitor fragment. (Control reactions at each end contain only test cDNA *or* competitor template.) (2) Analyse products on an agarose gel and define equivalence point. M, markers (size standards). (3) Compare the equivalence points obtained from different RNAs. The point at which the intensities of the competitor and cDNA bands are of equal strength is a measure of the specific RNA.

SUPPRESSION OF PROTEIN EXPRESSION

```
                              3' TACNNNN ...      Antisense oligo
                    Cap        | | | | | | |
mRNA          Gm⁷pppNNNNNNNNNNNAUGNNNNNNNN ........ NNNAAAAAAAAAA

            5' Untranslated   Translation      Coding       Poly A(3' tail)
                              initiation        region
                              codon
```

Fig. 3.16 One strategy for antisense suppression of protein expression – an (antisense) oligo annealed to the translation initiation site physically blocking the movement of the ribosome along the mRNA during translation. An oligo annealing elsewhere to the mRNA may also be effective in suppressing gene expression, either by blocking translation of the full length protein or by targeting the mRNA for digestion by the RNA/DNA hybrid-specific nuclease, RNase H.

Antisense suppresses gene expression by blocking translation of the mRNA or by destabilizing the mRNA.

copy because of the lower bond strength), or merely because it is shorter. Competitive PCR has the advantage of allowing comparisons to be made in the *plateau phase* of the amplification, where sufficient product has accumulated to allow easy detection. The products of regular PCR can only be compared in the exponential phase of amplification, which is tedious to determine and the products are difficult to detect, because the lesser template quickly 'catches up' and reduces any original differences.

Antisense gene suppression

Antisense suppression is where the expression of a protein is inhibited by the annealing of a matching (complementary) RNA or synthetic oligo to a sequence within the mRNA, thereby blocking or reducing the efficiency of translation (Fig. 3.16). Getting antisense RNA into the cell usually means first transfecting the gene in the reverse orientation (so that it is read 3'–5'), and then getting it expressed at high levels, both of which are difficult tasks. Since this method uses a large part of the gene in reverse orientation, it is also called antigene suppression.

Antisense oligonucleotides are much easier than antigenes to produce and test. Their small size and the ability to attain high concentrations (micromolar) extracellularly may be important for getting effective doses into the cell. However, experimental controls for what is often a very large amount of antisense oligo must include the sense oligo and non-sense oligo (the same nucleotides in a random order). Along with physically blocking translation, the annealed oligo may also function to reduce protein expression by rendering the mRNA susceptible to degradation by RNase H, which is specific for RNA/DNA hybrids (a heteroduplex).

FAILURE TO BLOCK PROTEIN EXPRESSION

Although antisense technology has shown promise in blocking the expression of some genes in tissue culture, many oligos that should bind the mRNA of interest fail to block expression of the protein. These failures may be due to the folding of RNA into secondary structures, such as hairpin loops, which exclude the oligo. Alternatively, it may be the oligo-induced stabilization of such secondary structures that inhibits translation. The imprecision of RNA secondary structure predictions and the inability to control the stringency of mRNA/DNA oligo annealing inside the cell (37°C and physiological salt conditions) may continue to limit the application of oligo-based antisense suppression techniques.

Peptide nucleic acids (PNAs)

PNAs are oligonucleotides that are joined together by peptide-like amide bonds instead of the phosphodiester bonds found in natural nucleic acids.

The binding of PNA to DNA is stronger than is DNA to DNA because their are no phosphates on the PNA to repel the phosphate backbone of the DNA and weaken the hybridization. Accordingly, the PNA–DNA or PNA–RNA binding is much less dependent upon the salt concentration of the solution, because there is no charge repulsion that requires shielding with higher ionic strength. PNAs are also relatively resistant to both nucleases and proteases. (Note that the term PNA, whilst helpful in conveying the concept, is actually a misnomer because the acidic portion of the DNA, the phosphate backbone, is not found in the PNA.) The increased strength of hybridization and stability may lead to a role for PNA in antisense inhibition of gene expression.

NUCLEIC ACID–PROTEIN INTERACTIONS

Overview

Gene expression is regulated by proteins that act as messengers, bringing the news of the world (or at least the cell surface) to the nucleus. The proteins activate or suppress gene expression by recognizing specific DNA sequences, called enhancers or repressors, near the gene (Fig. 3.17). The enhancer- or repressor-specific binding proteins regulate transcription of a gene by positively or negatively interacting with RNA Pol to increase or decrease transcription.

Several methods of analysing regulatory sequences are described here. A regulatory DNA sequence may be identified in

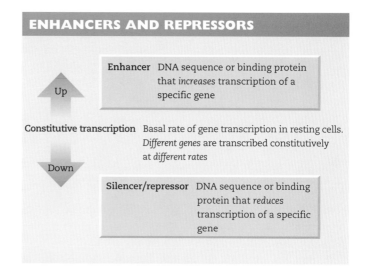

Fig. 3.17 Enhancers increase and repressors decrease the normal rate of gene transcription.

a reporter gene assay and the binding proteins analysed in the electromobility shift assay (EMSA) or footprinting binding assays. Finally, candidate binding factors (transcription factors) can be functionally tested in the reporter gene assay.

Reporter gene assay

Reporter genes are used to map DNA sequences *functionally* that regulate transcription.

Reporter genes are used to test whether pieces of DNA are involved in regulating the expression of the gene of interest. For example, a putative enhancer might be tested by placing a copy near a reporter gene (Fig. 3.18). If the piece of DNA increases expression of the reporter gene (i.e. it acts as an enhancer for the reporter gene) then it is assumed to function also as an enhancer for the original gene. Similarly, a sequence that is thought to stabilize mRNA might be tested by inserting it into the 3' UTR of a reporter gene and measuring its effect on the stability of the reporter mRNA. The important function of reporter genes is that they provide a test for an isolated part of a gene, so whatever effect is observed is probably caused by the fragment. However, some genetic elements probably only function in certain contexts, i.e. when they are together with many other elements, so the interpretation of reporter gene results can be difficult.

The *useful* aspect of reporter genes is that their products are easy to measure (Table 3.2). This is true because the gene of interest is replaced with something else, so it might as well be replaced by something with an easily assayed gene product.

ASSAY OF ENHANCER FUNCTION – 'REPORTER GENES'

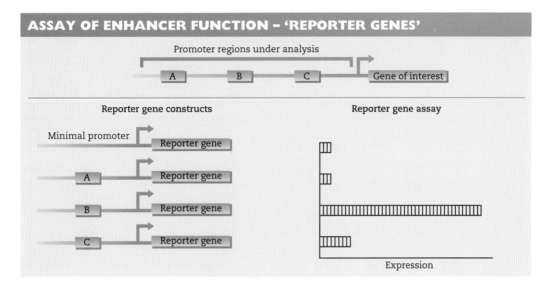

Fig. 3.18 Reporter gene assay of enhancer function in transfected cells. The minimal promoter (a TATA box and transcription initiation site) is inactive when transfected into a suitable recipient cell, so the reporter gene is not expressed and no product accumulates. Fragments of the promoter of the gene of interest are placed in front of the inactive, minimal promoter and these constructs are tested (fragments A,B and C). The expression of the reporter gene is greatly increased by the B region, suggesting that this piece of the promoter acts as an enhancer for the original gene of interest.

REPORTER GENE PRODUCT MEASUREMENT

Reporter gene	Assay method	Advantages/comments
Chloramphenicol acetyl transferase (CAT)	Radiochemical enzyme assay	Sensitive, many constructs already exist because it is *the old standard*
β-Gal	Chromogenic enzyme assay	Insensitive but cheap and safe, allows quantification of transfection efficiency
Luciferase (luc)	Luminescence (light production)	Very sensitive, near zero background in mammalian cells because it is a firefly gene
Human growth hormone (hGH)	Radioimmunoassay	Sensitive and a secreted product, so you just test the medium instead of lysing the cells

Table 3.2 Commenly used reporter genes

If the sequence of a regulatory element identified in a reporter gene assay resembles the sequence recognized by a previously characterized DNA binding protein, then that protein is a good candidate for regulating the reporter gene. The

ELECTROMOBILITY SHIFT ASSAY (EMSA)

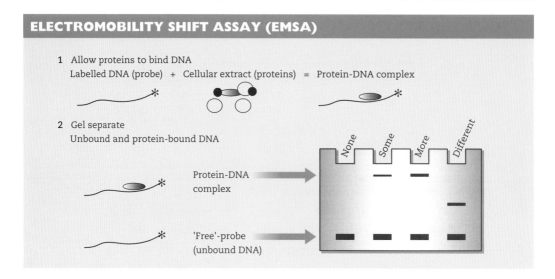

1 Allow proteins to bind DNA
Labelled DNA (probe) + Cellular extract (proteins) = Protein-DNA complex

2 Gel separate
Unbound and protein-bound DNA

Protein-DNA complex

'Free'-probe (unbound DNA)

Fig. 3.19 The EMSA of DNA-binding proteins, also known as 'gel-shift' assay. The protein-bound DNA moves more slowly than the free DNA through the gel (its mobility is reduced).

reporter gene assay can be reversed to test directly the putative transcription factor, instead of their binding site as illustrated in Fig. 3.18. A reporter gene under control of the regulatory DNA element is co-transfected along with an expression vector that produces the candidate transcription factor. If the transcription factor binds to the regulatory element and modulates transcription (as measured in the reporter gene assay), then the factor is likely to regulate the gene through binding to the DNA sequence.

EMSA/gel shift

The EMSA is used to detect specific nucleic acid-binding proteins. These proteins are potential regulators of gene expression.

DNA sequences act as enhancers or repressors by binding specific proteins. These proteins can be characterized in an electromobility shift assay (EMSA) in which fragments of the gene are labelled and incubated with proteins extracted from the nucleus (Fig. 3.19).

A modified form of this assay is used to identify by immunochemistry proteins binding to DNA sequences. An antibody that is specific for a previously identified DNA-binding protein is added to the incubation mixture (see Fig. 3.19, step 1).

Three things may be observed on the gel.

1 *No complex* is formed (the specific band disappears), probably because the antibody binds to the protein and interferes with the binding of the nucleic acid probe.

2 *No change* in the complex (the mobility of the band is not

DNASE I FOOTPRINTING

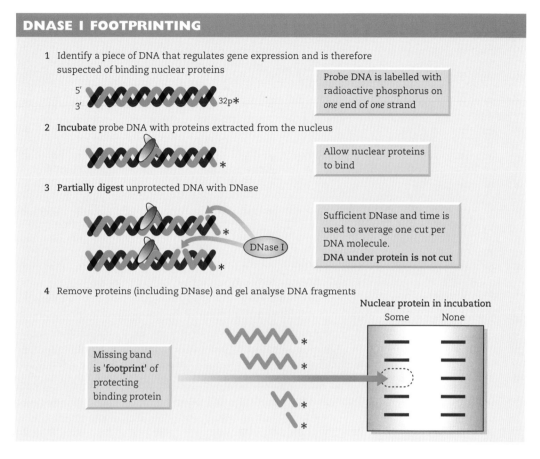

1 Identify a piece of DNA that regulates gene expression and is therefore suspected of binding nuclear proteins

Probe DNA is labelled with radioactive phosphorus on *one* end of *one* strand

2 **Incubate** probe DNA with proteins extracted from the nucleus

Allow nuclear proteins to bind

3 **Partially digest** unprotected DNA with DNase

Sufficient DNase and time is used to average one cut per DNA molecule.
DNA under protein is not cut

4 Remove proteins (including DNase) and gel analyse DNA fragments

Nuclear protein in incubation
Some   None

Missing band is **'footprint'** of protecting binding protein

Fig. 3.20 DNase I footprinting of DNA-binding protein. Without the protective DNA-binding proteins, all of the DNA would be digested. Note that well-separated cut sites are shown for purpose of illustration: DNase I can cut every unprotected phosphodiester bond (between nucleotides).

changed), probably because the antibody does not bind to the protein in the complex with the probe DNA.

3 A yet *larger complex* is formed (the original band is 'supershifted'), probably because the antibody binds to the DNA–protein complex. This larger complex moves more slowly through the gel.

Footprinting

Footprinting is used to detect and finely map specific nucleic acid-binding proteins. These proteins are potential regulators of gene expression.

Footprinting is another way to analyse DNA-binding proteins. It is a form of a nuclease protection assay in which the protector is a protein rather than an annealed oligonucleotide (Fig. 3.20).

EMSA VERSUS FOOTPRINTING

DNA-binding proteins are detected in both the DNase footprinting assay and the EMSA (described in the previous section), but there is an important difference between them. Whereas a (shifted) band can be seen in the EMSA even if only a small fraction of the probe DNA is bound, a large fraction of the probe DNA in the footprint assay must be bound and protected to be able to detect a missing band. This practical consideration may be one reason why the EMSA is used more often although footprinting can be more informative.

The footprinting assay reveals additional detail about the DNA–protein interaction, information that is not easily obtained from the EMSA.

Footprinting provides the following.

• Resolution of the *contact points* between the protein and the DNA at the nucleotide level. DNase is excluded from areas of close contact (yielding the footprint), so areas that are digested are unlikely to be involved in the binding.

• Information about the *DNA structure* in the DNA–protein complex. DNase I *hyper*sensitive sites are often observed alongside protected regions. Hypersensitive sites are thought to result from protein-induced bending of the DNA.

Finally, a modified footprinting procedure can determine whether a DNA–protein interaction observed *in vitro* occurs *in vivo*, inside the cell. This is important because the relatively large amounts of proteins and DNA that are used *in vitro* can produce misleading, artefactual binding. DNase I is not used in this assay because it does not pass through cellular or nuclear membranes. Instead, chemicals such as dimethylsulfoxide (DMS) are used because they are membrane permeable and react with DNA to form derivatives of nucleotides, as in DNA sequencing by the Maxam–Gilbert method. Living cells are treated briefly with DMS and then the genomic DNA is prepared. The DNA is cleaved at the derivatized nucleotides and the fragments are separated on a gel. The footprint, which is formed when binding proteins exclude DMS from reacting with the DNA, can be detected on a Southern blot or, for very little DNA, after PCR amplification.

Transgenic animals have genes (transgenes) added to their genomes. This allows one to see the effect of *new* or *differently regulated* genes.

TRANSGENIC ANIMALS (ADDING GENES)

Some questions cannot be answered with *in vitro* assays. Understanding the role of particular genes, especially those

TRANSGENIC ANIMALS

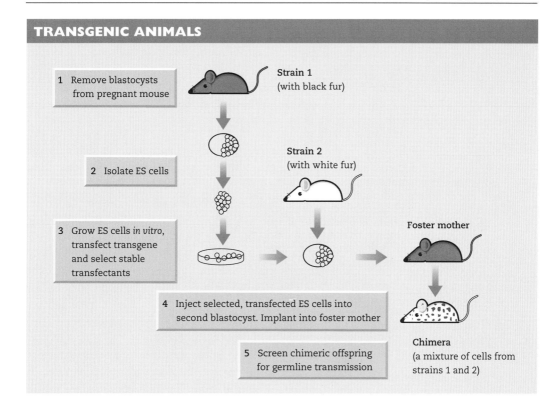

1 Remove blastocysts from pregnant mouse

Strain 1 (with black fur)

2 Isolate ES cells

Strain 2 (with white fur)

3 Grow ES cells *in vitro*, transfect transgene and select stable transfectants

Foster mother

4 Inject selected, transfected ES cells into second blastocyst. Implant into foster mother

5 Screen chimeric offspring for germline transmission

Chimera (a mixture of cells from strains 1 and 2)

Fig. 3.21 Transgenic animals – making an animal out of cells cultured *in vitro*. The cultured ES cells are added to a blastocyst and reimplanted into a foster mother. (Note that the foster mother contributes no genes to the chimera.) One formidable technical constraint has been that the ES cells can only be cultured for a limited time *in vitro* before they lose their ability to participate in the development of the animal.

involving development or the immune system, has benefited from the ability to place new genes back into animals. At some point, such technology may also allow the repair of disease-causing genes in humans.

Although the major aim is to cure or prevent debilitating diseases (an additional motivation is to understand more about life), transgenes have to date produced more sickness than health (although that tells us something about life). In transgenic mice, for example, overexpression of the normal cellular *myc* oncogene leads to adenocarcinomas, and expression of the viral SV40 transforming genes produces tumours.

The generation of transgenic mice followed the discovery that embryonic stem (ES) cells from a blastocyst (an early developmental stage, preceding implantation of the fertilized egg) could be isolated, grown *in vitro* and, when mixed with a second (fresh) blastocyst and placed in a foster mother, they would participate in the development of the mouse (Fig. 3.21). The resulting mouse is a chimera (mix) of the normal blastocyst cells and the cells grown *in vitro*. In some of the chimeras, the germ tissues develop from the ES cells. These animals pass on the transgene to their progeny.

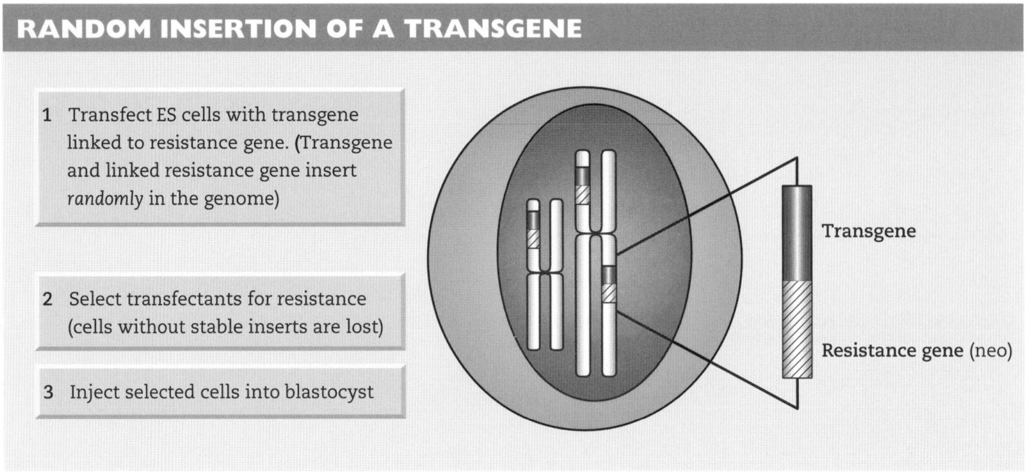

RANDOM INSERTION OF A TRANSGENE

1 Transfect ES cells with transgene linked to resistance gene. (Transgene and linked resistance gene insert *randomly* in the genome)

2 Select transfectants for resistance (cells without stable inserts are lost)

3 Inject selected cells into blastocyst

Transgene

Resistance gene (neo)

Fig. 3.22 Random insertion of a transgene into chromosomes of ES cells in preparation for making a transgenic animal. The ES cells are selected for stable integration of the transgene into the genome.

There are two types of transgenes.

1 Random insertions—where transgenes are randomly inserted all over the genome. Multiple copies are usually found in the transfected cell and chimera. Some of these transgenes may come under the control of different genes and be inappropriately regulated (Fig. 3.22).

2 Targeted, site-specific or homologous replacement are three names for the same event, the *replacement* of the original/normal gene with the transgene in the correct position in the genome. The frequency of homologous recombination is greatly increased by adding flanking DNA sequences that match the target region (Fig. 3.23). Nonetheless, these transgenes also insert randomly, so the subset of transfected cells containing the correctly inserted genes must be selected.

Generation of stable transfectants by random insertion of the transgene can be tedious and slow but it is much easier than the generation of site-specific transfectants. The principal difficulty with this method is in determining whether the randomly inserted transgene is inappropriately influenced by its new genetic neighbours. This can be evaluated by analysing several independently generated transfectants. Each independent transfectant is likely to have the transgene inserted into a different site in the genome, so the effects of different neighbouring genes are averaged out.

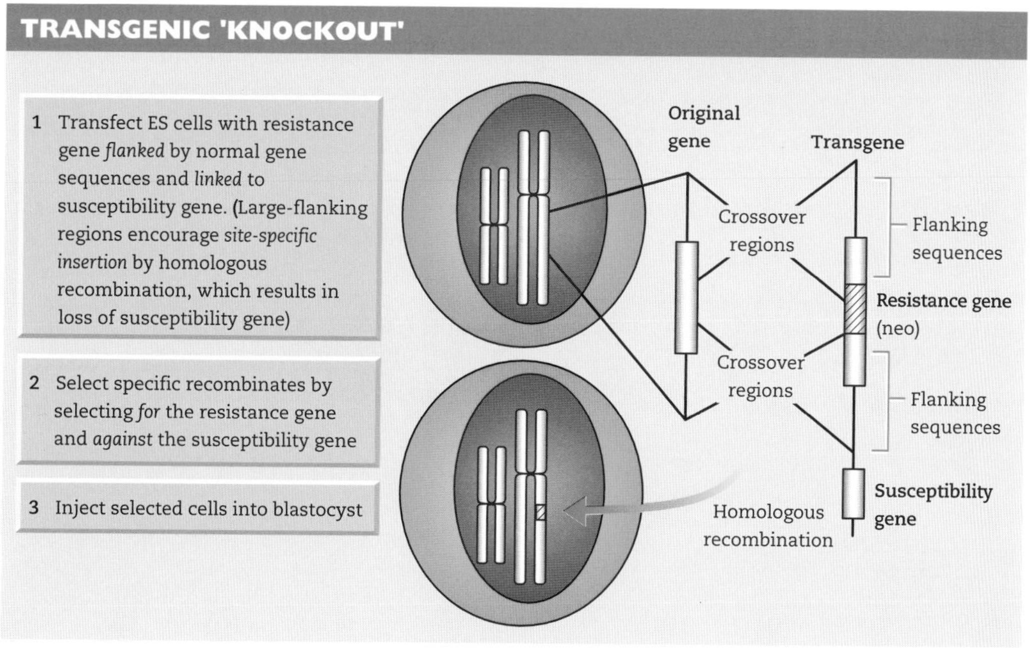

TRANSGENIC 'KNOCKOUT'

1   Transfect ES cells with resistance gene *flanked* by normal gene sequences and *linked* to susceptibility gene. (Large-flanking regions encourage *site-specific insertion* by homologous recombination, which results in loss of susceptibility gene)

2   Select specific recombinates by selecting *for* the resistance gene and *against* the susceptibility gene

3   Inject selected cells into blastocyst

Original gene

Transgene

Crossover regions

Flanking sequences

Resistance gene (neo)

Crossover regions

Flanking sequences

Susceptibility gene

Homologous recombination

Fig. 3.23 Transgenic 'knockout'. The resistance gene interrupts the gene that is targeted, resulting in a specific mutation of that gene. Note that although only one of the two genes has been mutated, a homozygous mouse could be bred by mating the heterozygous offspring.

In 'knockout' animals, a normal gene is replaced with a gene that does not encode the normal gene product.

KNOCKOUT ANIMALS (DELETING GENES)

Naturally occurring mutations that reduce or eliminate the activity of the encoded protein have long provided the first indication of the proteins' function. Recently, for example, the immune deficiency that results in elevated levels of the immunoglobulin M (hyper-IgM) provided the clue that the gene encoding the CD40 ligand is normally involved in the maturation of B cells, the producers of immunoglobulin. (Many additional examples are found in Chapter 4.) Gene 'knockout' technology can provide exactly the mutant you want now (well, soon anyway), without waiting for an 'experiment of nature'. Knockout mice have already helped to define the function (or often the apparent dispensability) of many genes.

Knockout mice are a subset of targeted transgenes, inserted by homologous recombination, *in which the target gene is interrupted by insertional mutagenesis* (Fig. 3.23).

A number of knockout mice are either apparently fine, which is surprising for the investigator but shows that the gene is not essential, or die as embryos with cause of death not easily determined. The most illuminating knockouts are those in which the deleted gene is important enough to cause a problem when it is missing, and the function can be inferred from the nature of the defect, but not so important that the knockout mouse dies too early.

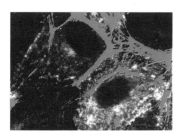

CHAPTER 4

Understanding genetics

EVOLUTION OF THE HUMAN GENOME

> There are only two amino acid differences between histone 2A in a pea and a cow.

The human genome contains DNA sequences which are closely related to those of other species. These DNA sequences appear to have been 'conserved' during the process of evolution. For example, genes encoding histone proteins are remarkably similar in different species.

In addition, genes with similar structure and function (gene families) occur throughout the human genome. This suggests

GENE FAMILIES

Gene families may be clustered together (as with the five β-like globin genes on chromosome 11) (Fig. 4.1), or dispersed throughout the genome (as with the ribosomal RNA (rRNA) genes of which over 300 copies are found on at least five different chromosomes)—some families stay close, others drift apart.

Histone genes are clustered at a few locations on chromosomes 1, 6 and 12.

Gene families often contain pseudogenes which are similar in structure to functional members of the gene family, but are themselves inactive because of alterations in either their regulatory sequences, coding regions, or both.

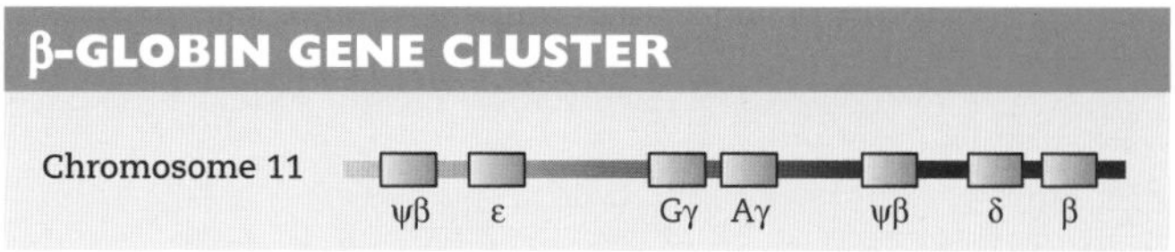

β-GLOBIN GENE CLUSTER

Chromosome 11

ψβ   ε   Gγ Aγ   ψβ   δ   β

Fig. 4.1 Map of the β-globin gene cluster. The different β-globin genes are ordered in the sequence in which they are expressed during development. ε-Globin is made very early in embryonic development, the two γ-globin variants are expressed during fetal life, and the δ- and β-globin genes take over after birth. Pseudogenes are indicated by ψ.

that the human genome has evolved from successive alterations, rearrangements and duplications of DNA.

The differentiated human cell contains 46 chromosomes of which two are sex chromosomes and the remaining 44 are autosomes. The female has two X sex chromosomes, and the male one X and a shorter Y sex chromosome. The autosomes are present in homologous (matched) pairs, so that each one has a partner with the same morphological appearance. Even though the human genome can be recognized as a distinct set of chromosomes with specific sizes and banding patterns, the DNA sequences and their organization are continually undergoing change, resulting in diversity between individuals. These alterations occur during the sexual process, and also arise from errors introduced into DNA (DNA mutations) during DNA replication, or as a result of environmental factors such as radiation or mutagenic chemicals.

MUTAGENIC CHEMICALS

Mutagenic chemicals include agents that are incorporated into DNA and distort its structure (such as acridine dyes), deaminating agents such as hydroxylamine and agents which hinder base pairing either by mimicking bases or adding alkyl groups (alkylating agents such as cyclophosphamide).

INHERITING GENES

The human somatic cell (any cell other than an egg or sperm cell) is said to contain 23 pairs, or a diploid number, of chromosomes (although X–Y is not strictly a pair). One chromosome of each pair is derived from the father and the other from the mother.

A new human being is created when egg and sperm (germ) cells unite to form a fertilized ovum or zygote. In preparation for fertilization the primitive germ cells undergo two meiotic divisions, during which the number of chromosomes is reduced to 23 or the haploid number (half the normal number).

First meiotic division

Before the first meiotic division the primitive germ cells replicate their DNA so that each chromosome is a double structure and the cell contains double the normal amount of DNA (Fig. 4.2). No further DNA replication occurs. During prophase, chromosomes condense so that each one contains two sister chromatids, and homologous chromosomes become intimately paired (with the exception of the X–Y combination).

Interchange of chromatid segments occurs at this stage through the process of *recombination* (Fig. 4.3).

RECOMBINATION

Recombination is the exchange of DNA between chromosomes. It is usually homologous–that is, it involves crossing over of similar DNA sequences, between a pair of chromosomes, each of which has condensed to form two chromatids. Breakpoints may occur within genes, in which case new fusion genes, each consisting of DNA sequences from both chromosomes will be formed. Unequal crossing over results in deletion of DNA sequences from one chromatid, and their duplication on the other. This creates variations in length, and may cause deletion of a gene from one chromatid and its duplication on another. Multigene families (see p. 87–88) are thought to have arisen through this process.

Non-homologous recombination, in which crossing over occurs between different sites either on the same or different chromosomes, is less common.

The chance of genes on the same chromosome passing to different germ cells as a result of recombination correlates roughly with the physical difference between the genes–the further apart the genes the greater the chance that they will be separated by recombination. The distance between two genes on the same chromosome can therefore be predicted according to how often the genes are separated by the process of recombination. The distance between two genes that show recombination in one out of 100 germ cells is one centimorgan (1 cM), which on average represents about one million base pairs (1 Mbp) of DNA. However, sites of recombinations are not random, and thus the actual physical distance represented by a centimorgan varies according to whether recombination occurs more or less commonly in the

MEIOSIS

Interphase Replication of chromosomal DNA occurs

Early prophase Homologous chromosomes approach each other and begin to pair

Middle prophase Homologous chromosomes condense to form chromatids, and pair along their entire length. Crossing over or interchange of DNA between chromatids occurs at this stage

Late prophase The intimately paired chromosomes begin to separate and move to the equator

Metaphase Chromosome pairs align on the equator

Early anaphase The chromosome pairs begin to separate. The chromatids in each chromosome remain attached

Late anaphase The chromosomes from each pair move to opposite poles of the cell

Telophase A nuclear membrane forms around each set of chromosomes

Interphase First meiotic division is complete. No further DNA replication occurs

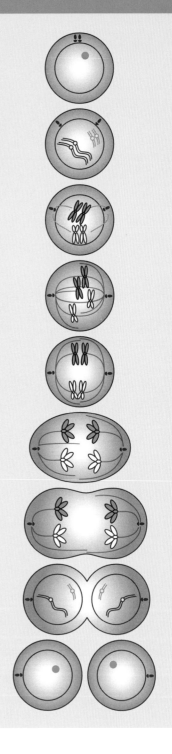

Fig. 4.2 First meiotic division.

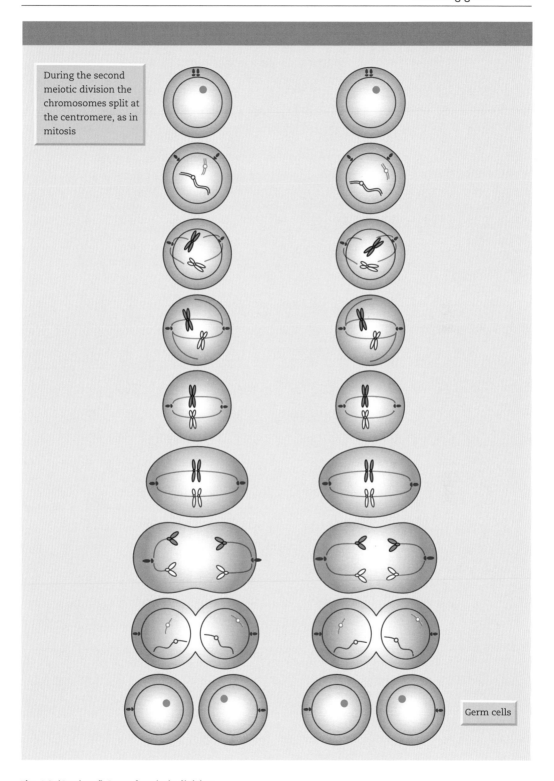

During the second meiotic division the chromosomes split at the centromere, as in mitosis

Germ cells

Fig. 4.2 (*Continued*) Second meiotic division.

RECOMBINATION

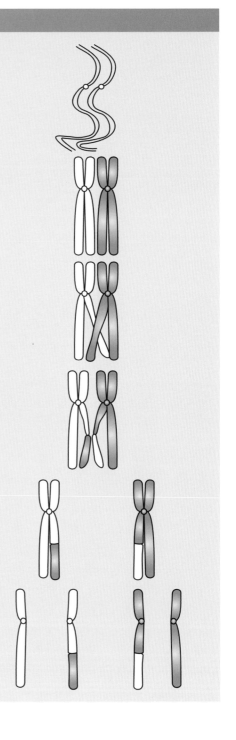

Homologous chromosomes which have replicated their DNA pair together

The chromosomes condense to form chromatids, and become closely aligned

Crossing over of intimately paired chromatids occurs

As the chromosomes separate, a break occurs in each chromatid at the point of crossing over, and there is an interchange of chromatid segments. More than one of these recombinations of genetic material may occur between each homologous pair of chromosomes

The chromosomes separate into two haploid cells at the end of the first meiotic division

The double-structured chromosomes separate during the second meiotic division to yield four germ cells, each of which is different

Fig. 4.3 Recombination of chromosomal DNA during meiosis.

region of interest. Recombination breakpoints appear to be particularly common within Alu sequences, raising the possibility that such sequences may be involved in promoting recombination.

ALU SEQUENCES

Alu sequences are short sequences of DNA which are recognized by the restriction enzyme Alu1. There are about 500 000 Alu sequences in the human genome. Although their function is unknown they may serve to promote recombination.

Different genes that are close together on a chromosome and therefore tend to assort together, rather than independently, are said to be *genetically linked*. This forms the basis for performing linkage analysis (see p. 105) to map genes in relation to each other.

Second meiotic division

After the first meiotic division each daughter cell contains one member of each chromosome pair. During the second meiotic division the paired chromosome strands divide at the centromere, so that each daughter cell receives 23 single chromosomes (Fig. 4.2).

Genes are inherited according to *Mendel's laws*.

• *Mendel's law of segregation*—each gene is represented in the genome by two versions (known as alleles, see p. 99), one being present on each of a pair of chromosomes. These two versions of the gene segregate during meiosis—they separate and pass to different egg or sperm cells.

• *Mendel's law of independent assortment*—genes at different loci pass to germ cells independently of each other.

Different genes on the same chromosome can still assort with different germ cells as a result of the exchange of DNA sequences between chromatids during meiosis (recombination).

DIVERSITY IN THE GENOME: WHY ARE WE DIFFERENT?

Diversity in the genome, therefore, arises during the sexual process—through the random assortment of chromosomes, and also *crossing over* or *recombination* of DNA between chromosomes. Alterations in genetic material also occur from DNA mutations.

Mutations in DNA may:
- have no effect on the expression of a gene;
- result in expression of a protein related but structurally different to the original;
- produce a DNA sequence which is untranslatable.

Mutations in promoter or enhancer sequences usually suppress gene expression.

Some diseases are always caused by the same mutation. In sickle cell anaemia (p. 98) a point mutation alters a single amino acid in the β-globin chain of haemoglobin. However, many different mutations in the α or β-globin genes give rise to thalassaemias (p. 125).

MANY DIFFERENT MUTATIONS CAUSE THALASSAEMIA

α-Thalassaemia is usually caused by large deletions in the α-globin gene, although a minority of cases are the result of point mutations, frameshift mutations or mutations in control regions of the gene. In β-thalassaemia many different point mutations or small deletions cause abnormal, reduced or absent β-globin production. Some mutations occur in control regions altering mRNA transcription, or splice sites altering mRNA processing. Major deletions in the β-globin genes are rare.

DNA mutations

Mutations in DNA can either be inherited or arise spontaneously. New mutations occur either in egg or sperm (germ) cells, or in differentiated tissues.

- Germ cell mutations do not lead to abnormalities in the affected individual, but can be inherited and cause disease in subsequent generations.
- Mutations in differentiated tissues are not inherited, but may lead to disease in the affected individual. Such mutations are of particular importance in the development of cancer.

DNA can be altered in various ways.

- *Point mutations*—substitution of one base for another.

POINT MUTATIONS

The most common mutation is substitution of cytosine (C) for thymine (T). C, when linked to guanine (CpG), is often methylated to give 5-methyl-cytosine which is unstable. Deamination of 5-methyl-cytosine yields T. CpG is therefore replaced by TpG.

DNA DELETION

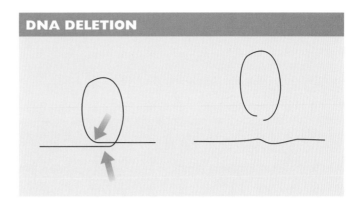

Fig. 4.4 DNA deletion: following breakage at the points arrowed the two ends of DNA reunite with loss of the loop.

Point mutations within coding regions may lead to incorrect amino acids being incorporated into the protein, or cause premature termination of transcription if a stop codon is introduced. Since the genetic code is degenerate (each amino acid is encoded by more than one base triplet), alteration of a single base may not alter the amino acid sequence. The best known example of a point mutation leading to an altered protein product is in sickle cell disease (see Fig. 4.9).

• *Deletion or addition of DNA sequences* (Fig. 4.4)—which alter both the length and sequence of DNA. They may arise as a result of unequal crossing over during meiosis, or spontaneous breakage of chromosomes.

The spontaneous chromosomal breakage rate is increased by ionizing radiation and mutagenic chemicals.

In some cases, this may involve the formation of a loop during DNA replication, which is lost allowing the two ends to reunite.

The size of deletions and insertions can range from a single base to several megabases. Large deletions result in the loss of whole genes or bits of chromosomes. Small deletions or insertions within coding regions can still cause havoc. For example, deletion of a single base will alter the remainder of the amino acid sequence (Figs 4.5 & 4.6).

Deleted fragments that lack a centromere are lost during subsequent cell division.

• *Translocation of DNA* (Fig. 4.7)—movement of DNA from one position to another involving recombination between non-homologous chromosomes during meiosis, or breakage of 2 chromosomes followed by abnormal repair. This does not usually result in loss of DNA. However, the expression of a gene can be dramatically altered by changing the surrounding DNA, thereby bringing it under control of different regulatory sequences.

• *Inversion of a DNA sequence* (Fig. 4.8)—two breaks occur in a chromosome, and the intervening segment is reinserted back-to-front. As with translocation, DNA is not usually lost, but genes may be disrupted at the breakpoints, or brought under control of different regulatory sequences.

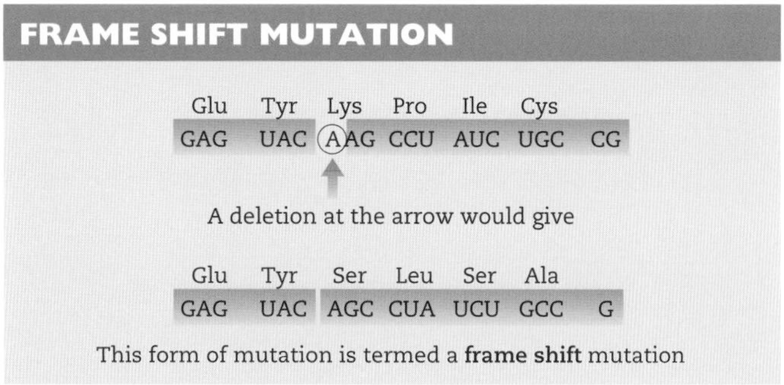

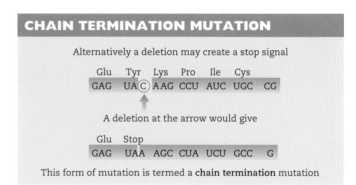

Fig. 4.5 A frame shift mutation.

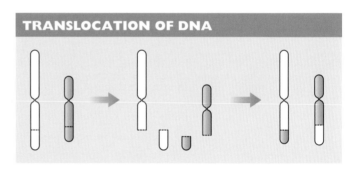

Fig. 4.6 A chain termination mutation.

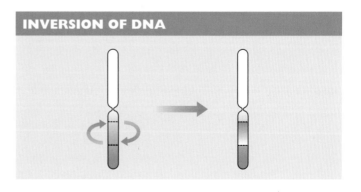

Fig. 4.7 Translocation of DNA.

Fig. 4.8 Inversion of DNA.

Trinucleotide repeat

A novel form of genetic mutation has recently been identified in which there is expansion of a sequence of DNA that contains a series of repeated nucleotide triplets. In diseases identified so far the repetitive sequence is present in the gene of normal individuals, but is expanded up to a 1000-fold in the gene of affected patients. Diseases which result from the amplification of trinucleotide repeats include:

- myotonic dystrophy—progressive muscle weakness in which there is continued muscle contraction of muscles after cessation of voluntary effort (see p. 127);
- fragile X syndrome—X-linked mental retardation (see p. 125);
- Huntington's chorea—progressive dementia and involuntary movements in middle age (see p. 126);
- X-linked spinobulbar muscular atrophy (Kennedy's disease)—muscle weakness associated with testicular atrophy and gynaecomastia;
- spinocerebellar ataxia type 1.

Trinucleotide repetitive sequences may occur in exons or untranslated regions of genes. The myotonic protein kinase gene which is linked to myotonic dystrophy normally has between five and 40 repeats of the trinucleotide CTG in its 3' untranslated region. In patients with myotonic dystrophy, up to 3000 CTG repeats have been found in the same region of the gene. In Kennedy's disease (X-linked spinobulbar muscular atrophy) there is amplification of a CAG repeat in the first exon of the androgen receptor gene.

The length of the trinucleotide repetitive sequence tends to increase as the gene passes from parent to offspring, providing an explanation for anticipation, the phenomenon by which the disease gets progressively more severe through successive generations.

Genetic variation: alleles and loci

Although each pair of chromosomes is morphologically similar, their exact genetic material varies as a result of chromosome recombination and random assortment during meiosis and DNA mutation. The rate of mutation varies throughout the genome. More mutations are found in non-coding regions (introns, gene-flanking regions, repetitive DNA sequences), than regions which code polypeptides (exons). This may re-

SICKLE CELL ANAEMIA

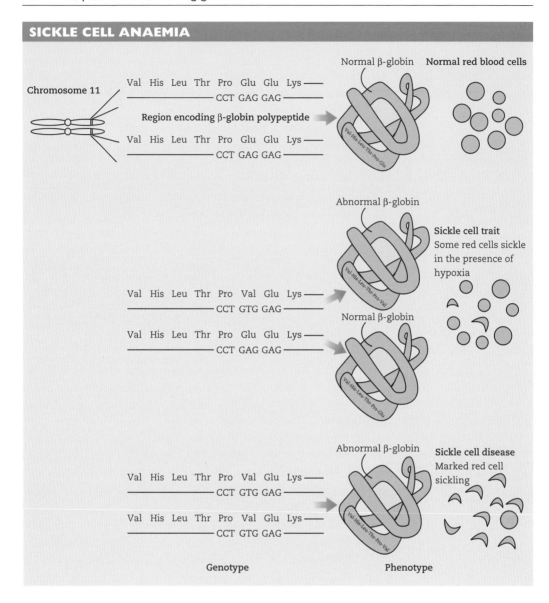

Fig. 4.9 Sickle cell anaemia.

flect the fact that *surviving* mutations are most likely to occur in 'unimportant' DNA (since most mutations are bad). The highest known mutation rate occurs in repetitive non-coding sequences such as minisatellite regions.

Variations in the DNA sequences of genes may (but need not) alter the protein which is encoded. For example, in sickle cell anaemia (Fig. 4.9) a single base substitution in the β-globin gene (GAG to GTG) leads to glutamic acid being replaced by valine in the β-globin chain of haemoglobin. The resulting haemoglobin molecule (haemoglobin S) differs in structure from normal haemoglobin and promotes sickling

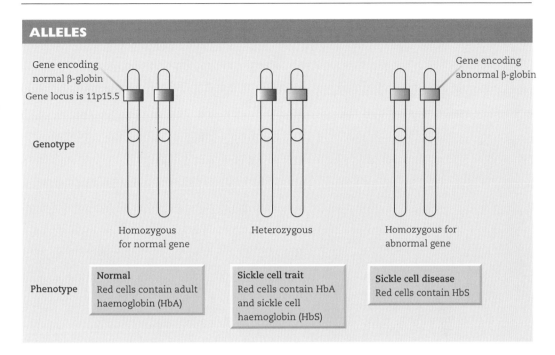

ALLELES

Gene encoding
normal β-globin

Gene locus is 11p15.5

Gene encoding
abnormal β-globin

Genotype

Homozygous
for normal gene

Heterozygous

Homozygous for
abnormal gene

Phenotype

Normal	Sickle cell trait	Sickle cell disease
Red cells contain adult haemoglobin (HbA)	Red cells contain HbA and sickle cell haemoglobin (HbS)	Red cells contain HbS

Fig. 4.10 Homozygous and hetero-zygous for the gene.

Blood group antigens are glycolipids and glycoproteins, and their expression depends on possession of the genes encoding the enzymes necessary for synthesis of the carbohydrate chains.

of red blood cells. However, alteration of GAG to GAA would still code for glutamic acid, and although the DNA sequence of the gene had changed, the protein product would be the same.

Alternatively, genetic variation may lead to production of different proteins, all of which are regarded as normal. For example, an individual's ABO blood group signifies whether they express A, B or O antigens on their red cells, any combination of which is considered normal.

A particular form of a gene is referred to as an *allele*, and the position of the gene on a chromosome is its *locus*. Thus, a pair of chromosomes may contain the same alleles at a given locus (the individual is *homozygous* for the gene), or the genetic material at a given locus may vary (the individual is *heterozygous* for the gene) (Fig. 4.10).

ALLELES

Each gene is present in two versions or alleles, one on each of a pair of homologous chromosomes. The two alleles may be identical or different, but both occupy the same position on the chromosome.

The occurrence of different alleles at the same site within the genome is known as *genetic polymorphism*. The physical characteristic resulting from expression of a single gene, or

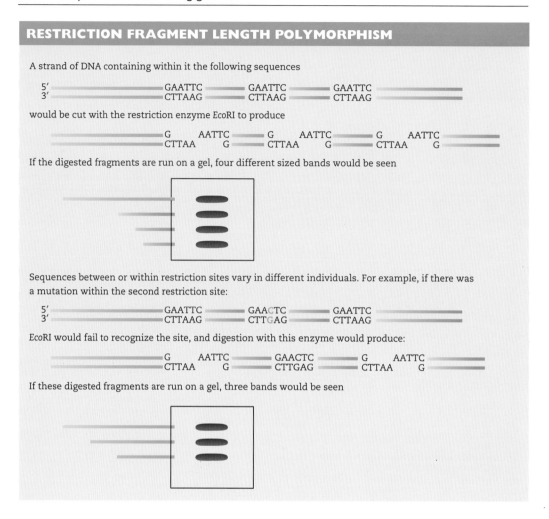

RESTRICTION FRAGMENT LENGTH POLYMORPHISM

A strand of DNA containing within it the following sequences

would be cut with the restriction enzyme EcoRI to produce

If the digested fragments are run on a gel, four different sized bands would be seen

Sequences between or within restriction sites vary in different individuals. For example, if there was a mutation within the second restriction site:

EcoRI would fail to recognize the site, and digestion with this enzyme would produce:

If these digested fragments are run on a gel, three bands would be seen

Fig. 4.11 Restriction fragment length polymorphism.

the interaction between several genes, is known as the *phenotype*.

A *dominant* gene manifests its phenotype if it is present on either one or both of a pair of homologous chromosomes. A *recessive* gene needs to be present on both chromosomes to influence the phenotype—a recessive phenotype is only manifest in homozygotes.

IDENTIFYING GENETIC DIFFERENCES BETWEEN INDIVIDUALS

Restriction fragment length polymorphism

A strand of DNA exposed to a *restriction enzyme* will be cut at specific sequences or *restriction sites* into a number of *restriction fragments* which vary in length. Alterations in the sequences within the restriction site, or variations in sequences between restriction sites, such as a change in the number of repetitive sequences in a non-coding region, will alter the length of the fragment. These alterations result in variations in the length of restriction fragments produced by cutting the same region of DNA from two individuals – there is 'restriction fragment length polymorphism' (RFLP) between the two individuals (Fig. 4.11). The variable length of such fragments can be demonstrated by separating them according to size in a gel.

Polymerase chain reaction (PCR) in genetic analysis

Once a region suspected of carrying a mutation has been identified, further analysis usually depends on the use of PCR to amplify large quantities of the DNA segment from normal and affected individuals.

Several methods can then be used to determine whether an individual carries a mutation.

• The primers can be designed so that one overlies the mutation. If the primer matches the normal DNA sequence, amplification will occur in normal individuals, whereas if the primer matches the mutated sequence, amplification occurs in affected individuals.

• The DNA segment can be amplified using primers that match sequences outside the region of the mutation, and the DNA analysed for the mutation using a separate procedure.

For example, the PCR products can be hybridized to labelled oligonucleotide probes which recognize either the normal or mutated sequence. Alternatively, if the mutation creates or destroys a restriction site, the amplified product can be examined to determine whether it is cleaved by the restriction enzyme.

Single-strand chain polymorphism

Single-stranded DNA tends to fold into a complex structure, which in part determines the mobility of the DNA strand in a non-denaturing gel. Even a single base change in the DNA can alter this conformation, and hence the mobility of the DNA sequence on a gel (Fig. 4.12).

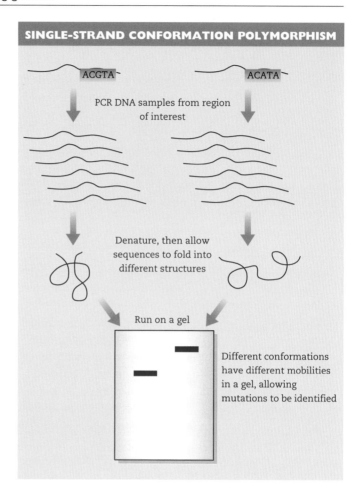

SINGLE-STRAND CONFORMATION POLYMORPHISM

ACGTA   ACATA

PCR DNA samples from region of interest

Denature, then allow sequences to fold into different structures

Run on a gel

Different conformations have different mobilities in a gel, allowing mutations to be identified

Fig. 4.12 Single-strand conformation polymorphism.

DNA fingerprinting

Although there is variability between genes of different individuals, most variation occurs in the non-coding regions of DNA. It is these regions, rather than genes themselves, that are used to create genetic finger-prints.

Non-coding DNA contains within it repetitive sequences that can be divided according to their size (Table 4.1).

• *Satellite DNA*—consists of arrays of repetitive DNA sequences that are not transcribed, and form the bulk of heterochromatin (chromatin that remains condensed during interphase). The size of the repeat unit varies from ~5 to 200 bp, and the satellite regions usually exceed 100 000 bp in length.

• *Minisatellite DNA*—regions are relatively short (9–70 bp), repetitive DNA sequences, that are scattered throughout the genome. Although variability occurs in the basic repeat unit,

SATELLITE DNA

Type	Size of repeated sequence (bp)	Size range of repeat unit	Features
Satellite	5–200	Big—may be several megabases	Found in heterochromatin and at centromeres
Minisatellite			
Telomeric family	6	1–20 kbp	Usually TTAGGG repeated about a thousand times. Protect the ends of chromosomes
Hypervariable family	10–60	1–20 kbp	Share a common core sequence GGGCAGGAXG (where X is any base you like)
Microsatellite	1–4	less than 1 kbp	Mononucleotide repeats of A (T on the complementary strand) and dinucleotide repeats of CA are the commonest. Collectively known as VNTRs they are interspersed throughout the genome

Table. **4.1** Characteristics of repetitive satellite DNA

most share a common core sequence. The number of repeat units at a given locus varies, and such regions are often referred to as variable number tandem repeats (VNTRs). Minisatellite regions are also found at telomeres (the ends of chromosomes), where they are usually formed by 6 bp repeated units.

• *Microsatellite DNA*—consists of runs of dinucleotide repeats, most commonly CA (or TG on the complementary strand), which occur throughout the genome.

Variations in the number of minisatellite repeats between individuals can be detected by using a restriction enzyme that cuts outside an array of repeats. These variations give rise to different sized DNA fragments, which can be separated in a gel and detected by hybridization with a probe that recognizes the repetitive unit. With the exception of identical twins, there are variations in minisatellite regions between all individuals, and these variable regions are inherited in a Mendelian fashion. This forms the basis for the technique of DNA fingerprinting, in which the size of minisatellite regions at numerous loci is determined by probing restriction fragments of total DNA (Fig. 4.13).

Linkage analysis investigates the possibility that a gene of interest is close to, and therefore often inherited with, a known genetic marker—either a restriction length polymorphism or another gene whose position has already been identified.

Recombinations occur through successive generations, and

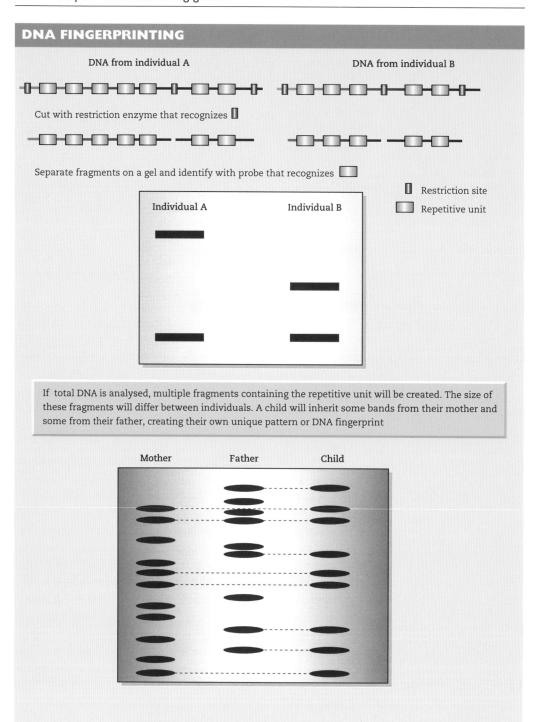

Fig. 4.13 Diagrammatic representation of DNA fingerprinting.

the chance of the gene and marker being inherited together is greatest if they are close together on the same chromosome.

LOD VALUES

The actual probability of a linkage is calculated using computerized statistical analysis, and is expressed as a *lod* (logarithm of the *odds*) score. Lod values >3 (representing a probability of 1000:1 as lod scores are to base 10) is generally taken as evidence of linkage.

FINDING GENES

The location of genes can be determined by using the techniques of genetic mapping (linking genes to other genes), and physical mapping in which the exact chromosomal location is defined. The exact approach depends on how much is known about the gene of interest.

If the DNA sequence is known, the gene can be physically mapped to a specific chromosomal region (*forward genetics*).

If the only information about a gene is the physical characteristic (phenotype) it produces, the gene may be shown to be close to a chromosomal region containing a known marker if it tends to be inherited with it (linkage analysis). The region of interest can then be examined for evidence of functional genes. Once a candidate gene has been identified it is cloned and sequenced. The amino acid sequence of the protein product can be deduced, and the cloned gene can be expressed in an artificial expression system. This process has been referred to as *reverse genetics*. Cloning a gene when nothing is known about it except its location is known as *positional cloning*.

Fluorescence in situ hybridization (FISH)

DNA sequences can be mapped directly to chromosomal regions by labelling a cloned DNA fragment and hybridizing it to a spread of metaphase chromosomes. Fluorescent labelling is most commonly used, and the technique is referred to as *FISH*.

Somatic cell hybridization

Another method for localizing genes to particular chromosomes is *somatic cell hybridization*. If human and rat cells are fused, the resulting hybrid cells randomly lose chromosomes,

CHROMOSOME WALKING

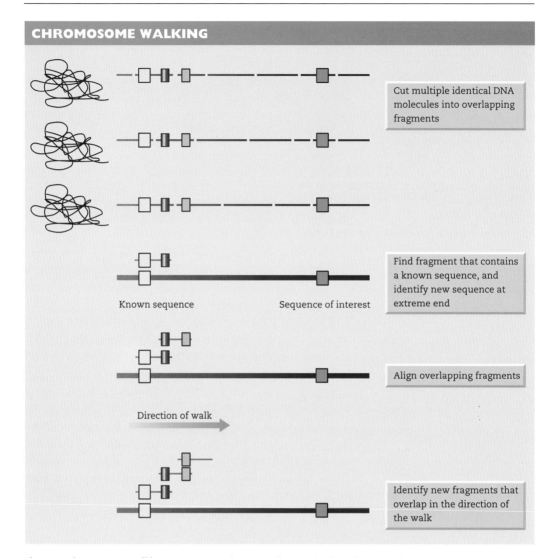

Cut multiple identical DNA molecules into overlapping fragments

Known sequence   Sequence of interest

Find fragment that contains a known sequence, and identify new sequence at extreme end

Align overlapping fragments

Direction of walk

Identify new fragments that overlap in the direction of the walk

Fig. 4.14 Chromosome walking.

producing cells in which only some human chromosomes are present. These cells can then be screened for the presence of either a known DNA sequence (using a labelled cloned DNA fragment), or the protein product of a specific gene.

Molecular gymnastics: chromosome walking and jumping

If a phenotype (usually a disease caused by an abnormality of the gene) can be linked to a marker with a known chromosomal location (a known gene or a restriction fragment length polymorphism), it is possible to explore that particular region by chromosome walking or jumping.

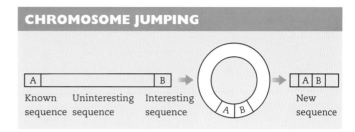

CHROMOSOME JUMPING

Known   Uninteresting   Interesting   New
sequence   sequence   sequence   sequence

Fig. 4.15 Chromosome jumping.

In *chromosome walking* (Fig. 4.14) a DNA fragment that contains the starting point (the marker which has been linked to a gene) is isolated from an appropriate library which contains overlapping regions of genomic DNA. The fragment is then used to find other cloned fragments that overlap with it. By repeating this procedure it is possible to 'walk' in steps created by each new fragment along the chromosome.

Chromosome walking involves characterizing every fragment from the starting point to potential candidate genes. Large intervening sequences can be skipped by *chromosome jumping* (Fig. 4.15) in which large fragments of DNA with a marker at one end are sized and then circularized so that the two ends are brought together. The circle is then itself cut into fragments, one of which now contains the marker linked to a sequence a measured distance away.

Potential sites of genes within large stretches of DNA can be identified by looking for *Hpa*II tiny fragments (*HTF*) *islands*.

> HTF islands are short stretches of DNA which are often found at the 5′ end of genes. They are rich in unmethylated CpG dinucleotides and can be readily identified as they are cut into tiny fragments by the restriction enzyme *Hpa*II, which has CCGG as its recognition site.

Initially, the techniques for mapping genes were limited by the size of DNA fragments that could be either separated by gel electrophoresis, or cloned into vectors. Large fragments of DNA can now be separated by *pulsed-field gel electrophoresis* (*PFGE*) (see p. 46), in which the resolution of fragments moving in a gel is improved by periodically changing the direction of the electric field. Limitations on the size of DNA fragments that can be cloned into vectors have been overcome by incorporating large fragments of human DNA into *yeast artificial chromosomes (YACs)*. Complete genomic DNA libraries can be created by fractionating human chromosomes and inserting fragments of up to 1 million bp in length into YACs. These DNA libraries can then be used to create physical maps of *contig*uous sequences (contigs) of human chromosomes.

Marking the map: sequence tagged sites

Reference points that can be easily identified are needed at numerous sites when constructing maps of the genome. Such landmarks are provided by sequence tagged sites. A sequence tagged site is simply a short sequence of DNA that can be amplified by PCR to give a unique marker for a particular region of DNA.

The Human Genome Project

A worldwide initiative to map and sequence all 3×10^9 bp of the human genome is now underway.

Extensive progress has already been made in both genetic and physical mapping. A physical map of the human genome constructed from a YAC library of overlapping cloned genomic DNA fragments was completed by Daniel Cohen and colleagues in 1993. The library contains 33 000 clones with an average insert size of 0.9 Mbp ($\sim$30 $\times$ 10^9 bp in total, or the equivalent of 10 haploid genomes). This presents a major step towards the full characterization of the human genome.

However, rather than investing all of the effort into sequencing large quantities of human DNA, emphasis has also been placed on efforts to improve sequencing technology, and to map and sequence the genomes of smaller model organisms. Such organisms merit study because they often have small, simple genomes, with a homologous gene pool to humans.

THE PUFFERFISH GENOME

The Pufferfish has a genome of only 4×10^8 bp, but contains a similar number of genes to humans. Recent mapping of this compact genome shows that it has small introns, little repetitive DNA, but conserved coding sequences with high homology to humans. Full characterization should therefore provide important information concerning coding regions of the human genome.

Sequencing of the *Haemophilus influenzae* Rd genome in 1995 represented the first complete genome to be sequenced from a free living organism.

The sequences of a genome represent only one step, albeit a large one, towards understanding how cells and organisms work. The enormous task of identifying genes and control regions within the sequence and determining their function will remain.

CHAPTER 5

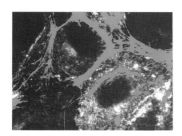

Molecular medicine in practice

In this chapter we will first consider the effect of recently developed technologies on clinical medicine. We will then describe how the understanding, diagnosis and treatment of a number of common clinical conditions have been influenced by advances in molecular medicine. These include monogenic disorders such as cystic fibrosis, polygenic disorders such as hypertension and diabetes mellitus, cancer, infectious diseases (including acquired immune deficiency syndrome (AIDS)) and the field of organ transplantation.

BIOTECHNOLOGY IN CLINICAL MEDICINE

Recombinant proteins

Perhaps the most widely recognized application of all of the recent advances in molecular medicine has been the use of recombinant DNA technology to make proteins. Once the gene for a particular protein has been cloned it can be inserted into the genome of microorganisms or cultured cells, which then produce the protein.

Recombinant DNA technology has allowed production of large quantities of numerous pure proteins which are now in use in clinical practice. The first recombinant human protein to be used clinically was the hormone insulin. Many others have followed (Table 5.1), including growth hormone, the thrombolytic enzyme tissue-type plasminogen activator (t-PA), erythropoietin (a hormone produced in the kidney that

RECOMBINANT PROTEINS IN CLINICAL PRACTICE

Recombinant protein	Therapeutic use	Year of product approval
Human insulin	Diabetes mellitus	1982
Human growth hormone	Growth hormone deficiency in children	1985
Interferon-α	Hairy cell leukaemia	1986
	Chronic hepatitis A and C	1992
Tissue-type plasminogen activator (t-PA)	Myocardial infarction	1987
Erythropoietin	Anaemia in chronic renal failure	1989
Granulocyte colony-stimulating factor (G-CSF)	Neutropenia following cancer chemotherapy	1991
Granulocyte–macrophage colony-stimulating factor (GM-CSF)	Myeloid reconstitution after bone marrow transplantation	1991
Factor VIII	Haemophilia A	1992

Table 5.1 Examples of recombinant proteins which have entered clinical practice

stimulates red cell production, and is deficient in patients with renal failure) and factor VIII (the clotting factor which is deficient in haemophilia). These recombinant proteins minimize the risk of transmitting infections which was encountered with proteins purified from human or animal tissue, and unlike animal proteins, are not recognized as foreign by the immune system.

The development of these products relies on considerable investment of both time and money from scientists and biotechnology companies, but the clinical and financial rewards are very clear.

Polymerase chain reaction (PCR)

PCR provides a simple, fast, sensitive method of amplifying minute quantities of DNA (and RNA).

Clinical applications of PCR include:
- the detection of bacterial, fungal or viral DNA (or RNA) in suspected infections (see p. 136);
- the amplification of segments of genomic DNA to look for mutations within known genes (see p. 101);
- tissue typing prior to organ transplantation (see p. 141).

In addition, the ability to amplify DNA from minute samples of tissue makes PCR an invaluable tool in forensic medicine.

Monoclonal antibodies

Antibodies are glycoproteins that are expressed on the surface of B lymphocytes. All of the antibodies on an individual B lymphocyte have identical antigen-binding sites, which are capable of recognizing a specific antigen. When an antibody on a B lymphocyte binds to an antigen, in the presence of factors released by T-helper lymphocytes, the B lymphocyte multiplies and differentiates into plasma cells which secrete antibodies bearing the same antigen-binding sites. Secreted antibodies bind to antigen, forming immune complexes. Immune complexes activate the complement system which can either directly destroy the foreign antigen or facilitate its recognition and phagocytosis by polymorphonuclear leucocytes or macrophages.

Antibodies are also known as immunoglobulins. Their basic structure is shown in Fig. 5.1.

Each antibody molecule has two identical heavy chains and two identical light chains. The amino acid sequences of the N-terminal domains vary between different antibody molecules, whereas the structure of the C-terminal domain is quite constant.

The N-terminal domain is known as the variable region, and is the region of the antibody that binds to antigen. Each antibody molecule has two antigen-binding sites. Most of the

IMMUNOGLOBULIN MOLECULE

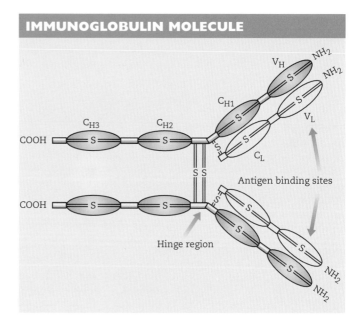

Fig. 5.1 The basic structure of an immunoglobulin molecule. The heavy chain (shaded) is made up of constant domains (C_{H1}, C_{H2}, C_{H3}) and a variable domain (V_H). The light chain has a constant domain (C_L) and variable domain (V_L). The structure is held together by disulphide bonds (=S=).

FIVE MAJOR CLASSES OF ANTIBODY

Antibody class (+subclasses)	Heavy chain	Amount in normal plasma	Characterisitic features
Immunoglobulin M IgM	μ	~10% of total	Forms large pentamers which are confined to the intravascular space. Appears early during immune response. Activates complement
IgG (IgG$_1$, IgG$_2$, IgG$_3$, IgG$_4$)	γ	Lots; ~75% of total	IgG$_1$ and IgG$_3$ efficiently activate complement. All subclasses cross the placenta
IgA (IgA$_1$, IgA$_2$)	α	~15% of total	Found predominantly as a dimer in mucosal secretions
IgD	δ	Very little; <1% of total	Found mainly on surface of circulating B lymphocytes
IgE	ε	Trace amounts only	Readily binds to receptors on basophils and mast cells

Table 5.2 Properties of human immunoglobulin classes

variation occurs in three hypervariable regions, each of which is only 6–10 amino acids long.

Tremendous diversity is needed in this region to be able to recognize the enormous number of pathogens it may encounter. It has been estimated that 100 billion different antibodies can be generated by 100 billion different lymphocytes.

The C-terminal domains of the heavy and light chains form the constant regions. The constant region of the light chain exists in two alternative forms, known as kappa ($\varkappa$) and lambda (λ). Antibody molecules have either two $\varkappa$- or two λ-light chains.

There are five major classes of antibody, which differ in both structure and function (Table 5.2).

Genetics of antibody production

The problem of how B cells are able to use their DNA to generate such an enormous diversity of antibodies caused quite a headache for immunologists, until it was discovered that the DNA encoding the different regions of an antibody molecule differs in adult B cells from that in embryo cells. It then became clear that a process of programmed rearrangement occurs in the chromosomal DNA of B cells as they mature. Different rearrangements bring particular combinations of coding DNA together, accounting for the tremendous variety of antibody molecules. The process is now clear to immunologists and a headache for students.

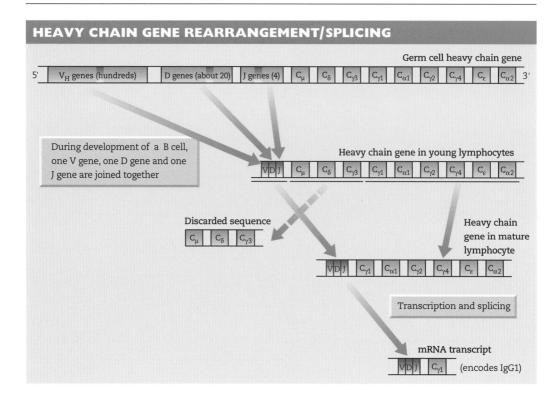

HEAVY CHAIN GENE REARRANGEMENT/SPLICING

Fig. 5.2 Rearrangement and splicing of heavy chain gene in a B lymphocyte destined to produce IgG$_1$.

Immunoglobulin (Ig) heavy chains, λ-light chains and χ-light chains are encoded in three different genetic loci on 14q, 22q and 2p, respectively. Each region differs in its organization, although each has multiple regions encoding potential variable sites in the final antibody molecule.

In germ cells the heavy chain gene is made up of about 300 variable (V) region genes, and a series of constant (C) region genes which encode the constant regions for IgM (Cμ), IgD (Cδ), IgG$_3$ (Cγ$_3$), IgG$_1$ (Cγ$_1$), IgA$_1$ (Cα$_1$), IgG$_2$ (Cγ$_2$), IgG$_4$ (Cγ$_4$), IgE (Cε) and IgA$_2$ (Cα$_2$). Between the V and C regions are two small coding regions, known as D and J. The D (for diversity) and J (for joining) regions each contain a number of D and J genes (Fig. 5.2).

During development of an individual B cell, a D segment is joined to a J segment, and the intervening DNA is deleted. The cell then selects one V-region gene that is joined to the preformed DJ segment. The product encodes the V region of the heavy chain, and is joined in the chromosome to all of the potential C regions. The cell then makes either IgM, IgD, IgG, IgA or IgE by a combination of further DNA rearrangement, and the joining of the V region domain to the relevant C region at the level of RNA processing.

A similar process occurs on different chromosomes in light chain genes, except that there is no D region in light chain genes. A V gene is joined to a J gene, and the product is joined at the RNA level to a gene encoding the C region of $\varkappa$- or λ-light chains.

The enormous diversity in antibodies synthesized by B cells results from the large number of V genes, and the way in which they can be combined with different D and J regions. Further diversity also occurs because the joining of V, D and J regions is imprecise.

Clonal nature of antibody production

The process by which antibodies are formed is known as *clonal selection*. Individuals produces an immense variety of B lymphocytes with different antibodies on their surface. A foreign antigen 'selects' from this population cells bearing antibodies that it can bind to. The antigen may be recognized by several different B cells bearing different antibodies, each of which may recognize a different site on the antigen. Each cell then proliferates into a large population of cells, all of which make the same antibody. A population of identical cells that has arisen from the same ancestral cell is known as a *clone*, and the single antibody that is produced by such a clone is known as a *monoclonal antibody*.

MYELOMA

Monoclonal antibodies can be found in the plasma of patients with *myeloma*. In this condition a clone of plasma cells undergoes malignant transformation, resulting in the production of a monoclonal antibody which can be detected as a discrete 'M' band on plasma-protein electrophoresis.

Production of monoclonal antibodies

The ability to produce and use monoclonal antibodies represents one of the most important contributions to molecular biology. Monoclonal antibodies provide a precise means of identifying and testing the function of specific proteins. These include known proteins which occur in health or disease, previously unknown proteins which can then be characterized and the products of recombinant DNA technology.

To produce a monoclonal antibody an immortal clone of B

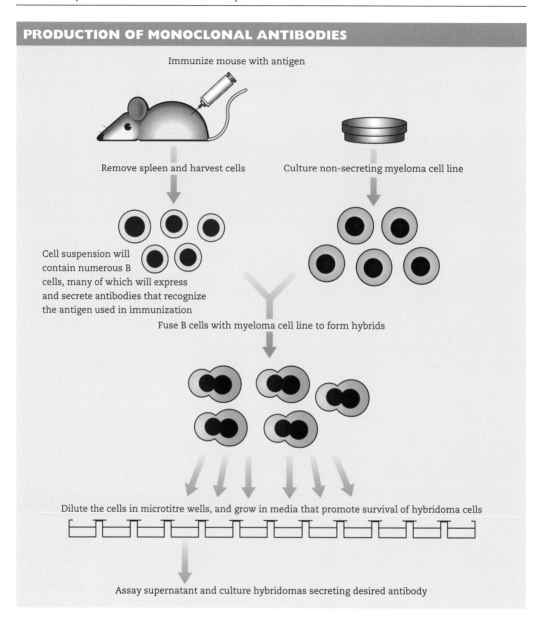

PRODUCTION OF MONOCLONAL ANTIBODIES

Immunize mouse with antigen

Remove spleen and harvest cells

Culture non-secreting myeloma cell line

Cell suspension will contain numerous B cells, many of which will express and secrete antibodies that recognize the antigen used in immunization

Fuse B cells with myeloma cell line to form hybrids

Dilute the cells in microtitre wells, and grow in media that promote survival of hybridoma cells

Assay supernatant and culture hybridomas secreting desired antibody

Fig. 5.3 Production of monoclonal antibodies.

cells is produced by fusing a B cell with an immortal malignant myeloma cell. The technology has been applied mostly to mouse antibodies, although mouse antibody genes can be modified so that the product resembles a human antibody.

In practice a mouse is injected with an antigen, and B lymphocytes that make an antibody to recognize the antigen are stimulated to grow and form clones in the spleen and bone marrow (Fig. 5.3). Spleen cell suspensions from the mouse will therefore contain B cells from a number of clones, which produce antibodies recognizing different sites (epitopes) on the antigen. These B-cell clones will not survive indefinitely in culture, but are made immortal by fusing them with a non-secreting myeloma cell line. The fused cells, which are known as hybridomas, produce a monoclonal antibody determined by the parent B cell, and have the immortality of the myeloma cell. Hybridomas are selected by growing the cells in media, which promotes their survival, and clones are screened by diluting the cells and assaying the supernatant for monoclonal antibodies of interest. Large-scale culture of the clones can then be used to produce quantities of pure monoclonal antibodies.

Applications of monoclonal antibodies

Monoclonal antibodies have found multiple applications in both basic science and clinical diagnosis and treatment. These are dealt with under the appropriate sections, but include the following.
• Identification of proteins in detection assays.
• Characterization of the structure and function of proteins using antibodies that specifically recognize different epitopes on a protein.
• Diagnosis of infectious diseases.
• Cancer diagnosis and treatment—by linking other molecules to antibodies that recognize tumour markers, they can be used to either image tumours or target drugs to tumours.
• Immunosuppression—antibodies directed against proteins expressed on lymphocytes can interfere with cell function. For example, the monoclonal antibody OKT3, which binds to the T-cell receptor complex, has been used as an immunosuppressant in transplantation. The usefulness of these antibodies is often limited because patients develop a human anti-mouse antibody response.

Genetic diseases

Inherited diseases result from a wide spectrum of genetic abnormalities, ranging from a single base change within a gene, as in sickle cell anaemia, to the addition or loss of a complete chromosome, as in Down's syndrome (Fig. 5.4) and Turner's syndrome (Fig. 5.5), respectively.

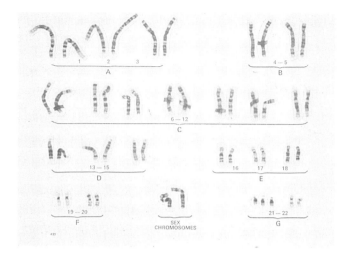

Fig. 5.4 Karyotype of Down's syndrome (trisomy 21).

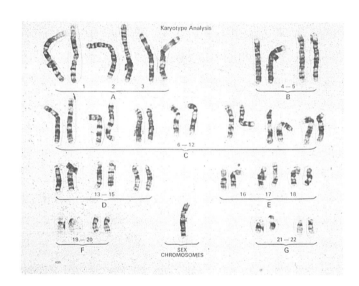

Fig. 5.5 Karyotype of Turner's syndrome (XO). Karyotypes courtesy of Genetics Laboratories, Addenbrooke's Hospital, Cambridge. Chromosomes can be precisely identified by banding patterns. In addition, chromosomes can be assorted into seven groups (A to G) according to size and shape.

TURNER'S AND DOWN'S SYNDROME

In Turner's syndrome only one X chromosome is present. Affected females are short and the neck may appear webbed. Ovaries fail to develop properly leading to primary amenorrhoea (menstruation never starts). IQ is normal.

In Down's syndrome an extra chromosome 21 (trisomy 21) results in the typical facial appearance (flat face, slanting eyes, small low-set ears) and Simian crease (single palmar crease), together with mental retardation and an increased incidence of congenital heart disease.

However, the most common human diseases are polygenic, resulting from the combined effects of multiple genes at different loci, each of which has a small but additive effect. Most of these also involve environmental factors, so that the cause is multifactorial.

Impact of molecular biology on genetic diseases

Once a gene has been fully characterized, abnormalities leading to diseases associated with it can be identified at the level of both the DNA sequences and the protein which it encodes. This often leads to an improved understanding of the disease process, and can allow detection of the abnormal gene in DNA samples from patients who: (i) have or are susceptible to the disease; (ii) are carriers; and (iii) the unborn fetus. The diagnosis of genetic disorders is only useful if it is of benefit to the patient or their family. Such benefits may be obvious if there is a readily available treatment for the condition. In many cases treatments are not available at present, and the options are restricted to genetic counselling (providing information and advice about the risk of transmitting an inherited disorder to offspring), or consideration of aborting an affected fetus.

Prenatal diagnosis

Screening for chromosomal abnormalities, such as trisomy 21 (Down's syndrome), is generally available for mothers at risk due to age or a previous history of chromosomal disorders. In addition, it is now possible to screen fetal DNA for the presence of specific gene disorders.

If the DNA sequence encoding the gene of interest is known, DNA from the fetus can be screened for the presence or absence of the normal gene. This usually involves developing a probe for the normal gene, and seeing if this recognizes a normal sequence in the fetal DNA, using the process of Southern blotting (see p. 68). An alternative approach is to determine whether a mutation has altered a restriction site within the gene, by looking for different sized fragments following digestion of the region of interest with a restriction enzyme.

SCREENING OF FETAL DNA

DNA for prenatal diagnosis is usually obtained from the fetus by amniocentesis or chorionic villus sampling.

Amniocentesis is performed between 14 and 17 weeks' gestation, and involves aspiration of amniotic fluid from around the fetus using a needle passed through the abdominal wall.

The chorion is a layer of fetal tissue that spreads over the uterine wall during early pregnancy. The chorion can be biopsied after the ninth week of gestation, either through the vagina and cervix or across the abdominal wall.

If the exact location and sequence of the gene is unknown, the probability of a fetus carrying the gene can be determined by linkage analysis. DNA from the fetus is screened for markers (other genes or restriction fragment length polymorphisms) which are known to be closely linked to the gene.

Gene therapy

The possibility that genetic disorders may be treated by correcting the underlying abnormality through gene therapy is now becoming a reality. Present efforts are focused on somatic gene therapy, in which defective genes are corrected in specific organs rather than egg or sperm cells. Two approaches have been used, *ex vivo* and *in vivo* gene delivery.

GENE DELIVERY

In *ex vivo* delivery, cells are taken from a patient, the new gene is inserted and the cells are then replaced. In *in vivo* delivery, the gene is targeted directly to the patient's tissues, usually by infecting them with a virus which contains the new gene.

Before considering the possibility of gene therapy it is necessary that:
• the gene must have been cloned and sequenced so that it is fully characterized and readily available in its correct form;
• it must be possible to introduce the gene safely and efficiently into appropriate target cells.

Any of the methods used to transfect cells can be used to introduce the gene, including direct injection, calcium microprecipitation and electroporation (see p. 60). However, one of the favoured methods is to incorporate the gene into a virus, which is then used to infect the target cell. Viruses have evolved to incorporate nucleic acid into cells and induce new gene expression, often without cytotoxicity. They are thus ideally suited to achieve a high efficiency of gene transfer and expression. The most commonly used viruses are derived from murine retroviruses. Although the use of viral vectors poses potential risks because of the pathogenic nature of viruses, most of the harmful viral genes can be removed.

Retroviral vectors

Retroviruses have a single-stranded RNA genome, which is converted into DNA within a cell by reverse transcriptase carried by the virus particle. The DNA is then incorporated into the host genome, where it is expressed to produce new viral RNA, together with proteins needed to form new viral particles. These are the *gag*, *pol* and *env* genes which code for the core proteins, the reverse transcriptase and the viral core proteins, respectively (see p. 138). Retroviruses can be prepared for gene therapy by replacing the *gag*, *pol* and *env* genes with the gene to be used for therapy.

One of the major advantages of retroviruses is their high efficiency of integration into the host genome, which is followed by stable expression of the introduced gene even after cell division. However, infection of non-replicating cells is poor, and the stability of the virus is adversely affected by the introduction of large gene inserts. Retroviruses are therefore best suited for the introduction of small genes into replicating cells.

Adenoviruses

Adenoviruses are double-stranded DNA viruses which have been used in live viral vaccines for many years, providing a long safety record. Adenoviruses can infect non-dividing cells, but viral DNA does not integrate efficiently into the host genome, and may not be transferred to daughter cells. This raises the possibility that in tissues undergoing cell turnover, introduced DNA may be lost, and need to be reintroduced at regular intervals.

Adenoviruses are particularly attractive for use in gene therapy of respiratory disorders because of the tendency of the virus to infect cells lining the airways.

Herpes simplex virus type 1 (HSV-1)

HSV-1 is a large double-stranded DNA virus which has been extensively studied as a human pathogen. Following primary infection (usually through mucous membranes of the mouth) it lies dormant in neurones, from where it may be reactivated, often causing 'cold sores' on the lips. The neurotrophic nature of this virus makes it a candidate for gene therapy in neurological disease.

HSV-1 vectors are similar in many respects to adenovirus vectors. They can infect non-dividing cells and are not integrated into the host genome. To make room for insertion of new genetic material, DNA sequences are deleted to render them replication defective, or to limit their toxicity.

Adeno-associated viruses
The adeno-associated virus is a small, non-pathogenic, single-stranded DNA parvovirus. It is incapable of autonomous replication, requiring co-infection with another virus (either adenovirus or herpes simplex virus). Because of its small size, adenovirus associated vectors can only accommodate inserts of less than 5000 bp. Adeno-associated viruses are unique in that they integrate into a specific site on chromosome 19.

The approach adopted when considering gene therapy varies considerably, and depends on the nature of the disorder and the organ involved. The haematopoietic system, the liver and skeletal muscle are amongst regions that are currently under investigation. Haematological disorders have provided the initial focus for gene therapy, as well-adopted procedures for bone marrow transplantation are already established. Bone marrow cells can be readily sampled, transfected and returned to the patient. In contrast, treatment of muscular dystrophy requires a different approach, as correction of the defective cytoskeletal protein dystrophin requires targeting of the new gene to a large number of muscle cells, making *in vivo* gene delivery the most feasible approach.

To illustrate the diverse nature of the problems encountered when considering gene therapy, examples of a number of genetic diseases that are currently undergoing trials of gene therapy, or are likely to become targets for such therapy in the future, are listed below.

Adenosine deaminase (ADA) deficiency
Clinical features   ADA catalyses the deamination of adenosine and deoxyadenosine. The gene is expressed predominantly in thymus and lymphoid tissue. Deficiency results in accumulation of deoxyadenosine and its metabolites,

which inhibits DNA synthesis. This leads to varying degrees of T- and B-cell dysfunction, including neonatal onset severe combined immunodeficiency (ADA–SCID), and milder, later onset immunodeficiency.

Disease-related gene ADA (20p).

Progress towards therapy Long-term stable expression of the ADA can be achieved ex vivo in human T lymphocytes using retroviral vectors. Significant immune reconstitution has been achieved in patients following periodic infusions with ADA gene corrected autologous T cells.

α_1-Antitrypsin deficiency

Clinical features α_1-Antitrypsin is the major serine–proteinase inhibitor present in blood. (Serine–proteinases are a group of proteolytic enzymes, including trypsin, chymotrypsin and elastase, in which the amino acid serine forms part of the active enzyme site). Deficiency of α_1-antitrypsin predisposes individuals to liver disease in childhood, and lung disease in adult life.

Disease-related gene α_1-Antitrypsin (14q) which normally encodes the 'M' subtype of the protein. Two common mutations give rise to 'S' and 'Z' variants of α_1-antitrypsin, in which abnormal folding of the protein alters both its structure (favouring formation of polymers which may become deposited in liver) and function.

Progress towards therapy Retroviral mediated transfer of the normal α_1-antitrypsin gene into cultured mammalian cells has been achieved. However, if the disease is in part related to the presence of the abnormal protein, transfer of the normal gene into patients may not correct the pathology completely.

Cystic fibrosis

Clinical features Autosomal recessive disorder of glandular tissue resulting in the production of abnormally thick secretions. Predominantly affects the respiratory tract and pancreas, leading to lung damage from recurrent chest infections and pancreatic insufficiency.

Disease-related gene Cystic fibrosis is caused by mutations of the cystic fibrosis transmembrane conductance regulator (CFTR) gene located on 7p. CFTR functions as a cyclic adenosine monophosphate (cAMP)-regulated chloride channel on the apical surface of airway and other epithelial cells.

Progress towards therapy   Transfer of the normal CFTR gene to cystic fibrosis epithelial cells *in vitro* corrects the defective chloride channel regulation. The feasibility of using an adenovirus vector to transfer and express CFTR cDNA in the respiratory epithelium of patients with cystic fibrosis has been demonstrated. This approach may therefore provide a strategy for correcting the cystic fibrosis phenotype.

Duchenne muscular dystrophy

Clinical features   X-linked recessive disorder leading to progressive skeletal and cardiac muscle dysfunction in children.

Disease-related gene   The cytoskeletal protein dystrophin (Xp) is either abnormal or absent. In Becker's muscular dystrophy a milder form of muscular dystrophy is caused by mutations of the dystrophin gene resulting in partially functional dystrophin protein which is reduced in amount or size.

Progress towards therapy   Problems are presented by the large number of affected muscle cells which need to be targeted, and the large size of the dystrophin gene (over 2 million base pairs), which is beyond the capacity of retroviral vectors. Truncated forms of the dystrophin gene encoding smaller, yet functional proteins (dystrophin minigenes) have been produced. This raises the possibility of gene transfer using viral vectors.

Familial hypercholesterolaemia

Clinical features   A deficiency of the receptor for low-density lipoprotein (LDL) results in hypercholesterolaemia (severe in the homozygous form, less marked in heterozygotes) leading to atherosclerosis. Patients die prematurely, usually from myocardial infarction.

Disease-related gene   LDL receptor (19p).

Progress towards therapy   LDL is processed in the liver, making hepatocytes the principal target for gene therapy. Replication-defective recombinant adenoviruses containing the LDL receptor gene have been expressed *ex vivo* in cultured hepatocytes from patients with LDL receptor deficiency. Preliminary reports indicate that cholesterol levels fall following insertion of infected hepatocytes into the liver via the portal vein.

Fragile X syndrome

Clinical features   X-linked mental retardation is associated with tall stature and a characteristic facial appearance.

Disease-related gene   The genetic defect is an expansion of a triplet repeat (CGG) in a gene of unknown function near the end of the long arm of the X chromosome (Xq27). The chromosome is constricted at this site, with partial detachment of the distal portion.

Progress towards therapy   Although the function of the gene is currently unknown, identification of the genetic defect offers the opportunity to detect female carriers, in addition to aiding prenatal diagnosis.

Gaucher's disease

Clinical features   Autosomal recessive lysosomal storage disorder in which the sphingolipid glucocerebroside accumulates in the liver, spleen and bone marrow. Low white and red cell counts, enlargement of the liver and spleen, and skeletal deformities occur.

Disease-related gene   Glucocerebridase (1q), which metabolizes glucocerebroside.

Progress towards therapy   Retroviral mediated transfer of the human glucocerebridase into cultured Gaucher bone marrow has been reported, raising the prospect of introducing transduced haematopoietic stem cells into patients.

Haemoglobinopathies

Clinical features   Normal adult haemoglobin is made up of two polypeptide chains, the α- and β-chains, which are folded such that each chain can hold an oxygen-binding haem molecule. The haemoglobinopathies are a diverse group of autosomal recessive disorders of haemoglobin synthesis which include sickle cell anaemia (abnormal β-chain synthesis) and the thalassaemias (deficient or absent α- or β-chain synthesis). Together they form the most common group of single gene disorders in the world population.

Disease-related genes   Genes encoding five different β-globin chains and three different α-globin chains are expressed in a precisely regulated manner during different stages of development. For example, during fetal life the two β-globin variants called γ-globin, combine with two α-globin chains to give rise to fetal haemoglobin. During adult life the β-globin variants themselves combine with α-globin chains to form adult haemoglobin. The five β-globin chain genes are clustered on chromosome 11, whereas the α-globin chain genes occur together on chromosome 16. Numerous different mutations in the α-globin and β-globin genes have been described, which give rise to α- or β-thalassaemia, respectively. Sickle cell anaemia is caused by a point mutation, which involves substitution of T for A in the second nucleotide of the sixth codon changing the sixth amino acid from glutamine to valine.

Progress towards therapy   Globin genes are highly expressed only in erythroid cells in a tightly regulated manner. Attempts to produce high level regulated expression of globin genes have so far been unsuccessful.

Haemophilia

Clinical features   Sex-linked recessive clotting disorder in which patients suffer mainly from spontaneous bleeding into joints and soft tissues, and excessive bleeding in response to trauma or surgery.

Disease-related gene   Haemophilia A (classical haemophilia) or haemophilia B (Christmas disease) result from defects in the clotting factor VIII (Xq28) or factor IX (Xq27), respectively.

Progress towards therapy   The lack of requirement for tissue-specific expression, or precise regulation of the deficient factors (small amounts have significant clinical benefits and large amounts do not appear harmful), make haemophilia an excellent candidate for gene therapy. Initial studies have introduced murine myoblasts transduced by a retroviral vector with human factor IX into mice, and found detectable circulating factor IX.

Huntington's disease

Clinical features   Progressive dementia and involuntary movements (chorea) in middle age. Inheritance is autosomal dominant, although the disease is late and variable in its presentation.

Disease-related gene   The Huntington's disease gene (4p16) contains an expanded CAG trinucleotide repeat.

Progress towards therapy   The gene encodes a novel protein of currently unknown function. Identification of the genetic mutation increases the precision of genetic testing. However, presymptomatic testing for Huntington's disease must be approached with great care because of the implications of a positive test for the patient and family.

Leucocyte adhesion deficiency

Clinical features Autosomal recessive disorder characterized by recurrent bacterial infections, impaired wound healing and impaired pus formation.

Disease-related gene CD18 gene (21) encodes a subunit of the leucocyte function associated antigen 1 (LFA-1). LFA-1 interacts with another cell surface protein, intercellular adhesion molecule 1 (ICAM-1), which is expressed at high levels in inflamed tissues.

Progress towards therapy Bone marrow cells from CD18-deficient patients infected with a retrovirus encoding the CD18 gene, express CD18 in long-term culture.

Myotonic dystrophy

Clinical features Progressive muscle weakness in which there is continued muscle contraction of muscles after cessation of voluntary effort (myotonia). The disorder is autosomal dominant and may be associated with cataracts, frontal baldness, mild intellectual impairment and cardiomyopathy.

Disease-related gene There is expansion of a CTG repeat in the 3' untranslated region of the myotonin protein kinase gene (19q13).

Progress towards therapy Identification of the genetic defect has increased both the potential for screening and our understanding of the disease. The degree of expansion correlates with the severity of the disease, and anticipation is characteristic (increasing severity of the disease in successive generations due to progressive expansion of the repetitive unit).

Polycystic kidney disease (adult)

Clinical features Autosomal dominant condition in which renal failure results from progressive cystic degeneration of the kidneys.

Disease-related gene Approximately 85% of cases are due to a defect in Polycystic kidney disease 1 (PKD1). The gene was identified by analysis of a Portuguese family in which tuberous sclerosis (mapped in 1993 to 16p13.3) and adult polycystic kidney disease exist in different family members. A mother and daughter with adult polycystic disease (but not tuberous sclerosis) were found to have a balanced translocation between chromosomes 16 and 22, with a breakpoint on chromosome 16 at 16p13.3. The breakpoint disrupted a gene encoding a 14 kbp transcript, and the identification of different mutations of that gene in other patients with adult polycystic kidney disease confirmed that it was PKD1. Even though the gene has been identified, isolation and characterization of the full length transcript has proved difficult. This is in part because more than three-quarters of the gene is duplicated elsewhere on chromosome 16. Thus, complementary DNA (cDNA) libraries contain sequences of the duplicated transcripts as well as the PKD1 transcript. Screening for mutations also requires care to ensure that they arise in PKD1 rather than duplicated regions.

The PKD2 gene, responsible for most non-16p linked polycystic kidney disease has been localized to 4q13–q23.

Progress towards therapy   The underlying cellular defect is at present unknown. Analysis of the sequence available so far has not revealed homologies to other known proteins. Characterization of the full protein sequence should help to clarify the molecular pathology.

Tuberous sclerosis

Clinical features   Autosomal dominant multisystem disorder in which the principal features are adenoma sebaceum (red nodules on the face), epilepsy and mental retardation.

Disease-related gene   Loci have been assigned to chromosomes 9 and 16. The chromosome 16 gene, designated TSC2, encodes tuberin.

Progress towards therapy   Recent identification of TSC2 increases the potential for gene therapy.

Wiskott–Aldrich syndrome

Clinical features   Eczema, immunodeficiency and thrombocytopenia in affected males.

Disease-related gene   Abnormal gene has been localized to Xp, but the protein product is currently unknown.

Progress towards therapy   Gene is thought to be expressed primarily in bone marrow cells raising potential for gene therapy.

Polygenic disorders

The nature of these diseases inevitably makes their genetic analysis complex. However, advances in physical mapping of the genome have facilitated identification of different susceptibility genes which are likely to be involved in the development of the same disease.

Three principal strategies have been used to identify susceptibility genes in polygenic disorders:

• *linkage analysis* — tests for segregation of traits with genetic markers beyond that expected by chance in families, and is applied to affected siblings or relative pairs to increase the power of detection;

• *association studies* — the comparison is between unrelated cases and controls;

• *candidate genes* — can be directly analysed for mutations.

CANDIDATE GENES

Candidate genes are genes that might be expected to be involved in the development of a multifactorial disease. For example, the genes involved in lipid metabolism are important candidates in trying to understand the polygenic inheritance of cardiovascular disease.

Consideration of recent advances in the mapping of loci for diabetes mellitus and hypertension highlight both the achievements, and the potential difficulties, in unravelling the complexities of polygenic diseases.

Diabetes mellitus

Insulin-dependent diabetes mellitus (IDDM; type I diabetes) is an autoimmune disorder in which the insulin-producing ,-cells of the pancreas are destroyed. The identical twin of an IDDM patient has a 30–50% chance of developing the disease, implying that both genetic and environmental factors are involved.

Two chromosome regions have been established as being associated with, and linked to, IDDM: (i) the major histocompatibility complex (MHC) class II region (designated IDDM1; 6p21); and (ii) the insulin gene region (IDDM2; 11p15). The recent construction of high-resolution human genetic linkage maps has allowed a total genome screen in sibling pairs with type I diabetes to identify susceptibility genes. This tremendous achievement has allowed identification of 18 other regions showing positive evidence of linkage to type I diabetes. These include regions on 15q (IDDM3), 11q (IDDM4) and 6q (IDDM5). The major locus is IDDM1, but the inheritance of diabetes is clearly complex and polygenic.

Hypertension

The identification of susceptibility loci in hypertension has proved more challenging. Blood pressure levels show strong familial aggregation which cannot be accounted for by shared environment alone. However, the genetic and environmental factors contributing to hypertension are likely to be extremely diverse, confounding the search for responsible genes. Attention has principally been directed towards the identification of candidate genes. Initial results suggest that polymorphisms in renin, angiotensin converting enzyme nor Na/H antiporter genes, the products of which would be expected to influence blood pressure, do not commonly contribute to human hypertension.

Cancer

Cancer usually arises as a result of acquired genetic changes. Environmental factors that can damage DNA, including radiation, viruses and certain chemicals (see p. 88), are therefore important, and in some cases the susceptibility to such changes is inherited. A number of genes that are important in the development of tumours have been identified. It seems likely that several different genetic alterations are necessary to produce most human cancers.

Oncogenes

Retroviruses transcribe their RNA genome into DNA, which then becomes inserted into the host genome. Certain retroviruses can cause tumours in animals, and many of these 'tumour viruses' contain genes that are capable of inducing malignant transformation in cells. These genes are called *viral oncogenes* or *v-onc genes*. Sequences that are homologous to *v-onc* genes are normally found in the genome of all vertebrate species. Thus, each *v-onc* gene that has been identified has a normal cellular counterpart in the human genome, which is known as a *cellular oncogene* (*c-onc gene*) or *proto-oncogene*.

There are now numerous examples of these *c-onc* genes that have important regulatory functions in normal cell division and differentiation.

• *v-fos* causes osteosarcoma in mice. *c-fos*, a homologous sequence on the long arm of chromosome 14 encodes a protein which binds to DNA and regulates the expression of a number of genes.

• *v-erb-B* causes erythroblastosis in chicken. *c-erb-B* on the short arm of chromosome 7 encodes a truncated form of the receptor for epidermal growth factor, which appears to signal even in the absence of the growth factor.

• The *ras* family of retrovirus oncogenes cause sarcoma in the rat. Homologous sequences in the human genome encode GTP-binding proteins which are involved in cell signalling.

Oncogenes are therefore normal cellular genes which encode proteins involved in normal cell growth and division.

Tumour-suppressor genes

Another group of genes involved in the development of cancer are tumour-suppressor genes or *anti-oncogenes*. The presence of these genes in normal cells is thought to suppress the development of tumours. Mutation of the gene results in loss of suppression, favouring malignant transformation. Examples are the *p53* gene and the retinoblastoma gene. The *p53* gene, on the short arm of chromosome 17, encodes a DNA-binding protein of molecular weight 53 kDa, which suppresses tumour cell growth.

How do oncogenes cause cancer?
Oncogenes are usually involved in the regulation of cell growth and death. Disruption of their normal function can occur through a number of processes. Deletion of a tumour-suppressor gene results in the loss of its function, and mutations within a gene may lead to production of an abnormal protein product. Gene expression may be increased by amplification of the gene, so that it is present in multiple copies, whereas translocation from one chromosomal location to another may disrupt the gene, or bring it under the control of different promoter regions.

A number of consistent chromosomal translocations involving oncogenes have been described in human malignancies.

Burkitt's lymphoma. In the majority of patients with this lymphoid tumour there is a translocation between chromosomes 8 and 14. This results in the *c-myc* gene (8q34), which is an important regulator of cell growth and cell death, being juxtaposed to the immunoglobulin heavy chain locus (14q32), thereby activating the oncogene.

Philadelphia chromosome. Reciprocal exchange of chromosomal material between chromosomes 9 and 22 gives rise to the Philadelphia chromosome in the malignant cells of patients with chronic myeloid leukaemia. The rearrangement results in the translocation of the *bcr* (breakpoint cluster region) gene (22q11) adjacent to *c-abl* (Abelson murine leukaemia virus) (9q34), resulting in a fusion gene, and the subsequent expression of the BCR–ABL fusion protein which is involved in the malignant transformation of myeloid cells.

Too much growing or not enough dying?

The precise cellular mechanisms by which genetic alterations cause cancer are at present unknown. Recent attention has challenged the concept that cancer arises due to the abnormal proliferation of cells, and has focused on the idea that malignant cells escape the normal process by which cells die.

APOPTOSIS

The lifespan of cells is normally controlled by a physiological process of programmed cell death, known as *apoptosis*.

   Apoptosis is characterized by the condensation of cytoplasm and chromatin, and the fragmentation of nuclear DNA such that it appears as a 'ladder' of different sized fragments when run on a gel. The cells disintegrate into apoptotic bodies, and are rapidly eaten by neighbouring cells.

The development and progression of certain malignancies may result from the inability of cells to die, and the characterization of genes involved in apoptosis has provided insights into the molecular mechanisms of oncogenesis. Furthermore, it is becoming increasingly clear that many cytotoxic drugs used in the chemotherapy of cancer work by inducing apoptosis in tumour cells.

The *c-myc* encoded protein is a DNA-binding transcription factor that paradoxically can induce both proliferation and apoptosis in cells. Whether a cell grows or dies in response to *c-myc* is likely to depend on the availability of other critical factors. For example, the insulin growth factor 1 (IGF-1) suppresses *c-myc*-induced apoptosis, whereas cytotoxic drugs promote apoptosis. Deregulation of *c-myc* expression appears to be fundamental to the development of a number of malignancies.

The *bcl-2* oncogene (located on 18q21) encodes a protein that localizes to the inner mitochondrial membrane, and has also been detected in the nuclear envelope. The *bcl-2* encoded protein blocks apoptosis and promotes cell survival.

Evidence for a role of *bcl-2* in carcinogenesis comes from genetic studies of patients with tumours of the lymphoid system. Reciprocal translocation between the long arms of chromosomes 14 and 18 is seen in 85% of patients with follicular lymphoma. In this translocation the *bcl-2* gene is juxtaposed to the immunoglobulin heavy chain gene, resulting in greatly enhanced expression of the bcl-2 protein.

Although abnormalities of the *p53* tumour-suppressor gene are frequently encountered in tumours, the extent to which *p53* is involved in regulating apoptosis is at present unknown.

Familial cancer

Sometimes, the development of cancer depends on a mutation occurring during life in a particular form of a gene, which has itself been inherited. Most familial cancers involve defects in tumour-suppressor genes.

Retinoblastoma

Retinoblastoma is a malignant tumour of the retina affecting about 1 in 20000 children. About 25% of cases are familial, and are associated with mutation or deletion of a 'tumour-suppressor'·gene on the long arm of chromosome 13. Familial cases are dominantly inherited, so if a mutation is inherited on one chromosome, a sporadic mutation must occur on the other chromosome. A wide range of mutations occur with a relatively high frequency, perhaps as a consequence of the large size of the retinoblastoma gene (>200 kbp). Sporadic cases require that two separate mutations occur on both copies of chromosome 13.

Wilm's tumour

Wilm's tumour is an embryonal malignancy of the kidney that affects about 1 in 10000 children. One Wilm's tumour-suppressor gene (WT-1) has been located (11p13), and found to encode a transcription factor that is critical to normal kidney and gonadal development. A second Wilm's tumour-suppressor gene has been identified at 11p15, and linkage studies suggest that further loci may exist.

Breast cancer

A small proportion of breast cancer cases, estimated at 4–5%, are due to highly penetrant dominant genes. A number of genes have been linked to familial breast cancer.

• *BRCA1* — thought to account for ~45% of families with an increased incidence of breast cancer, and at least 80% of families with an increased incidence of both breast cancer and epithelial ovarian cancer.

A strong candidate for the *BRCA1* gene has been mapped to 17q21 by positional cloning. A number of different mutations in the gene have been identified in kindreds affected by breast cancer at young ages. *BRCA1* appears to encode a tumour-suppressor protein that acts as a negative regulator of tumour cell growth. The predicted protein has 1863 amino acids and contains a zinc finger domain, similar to those found in DNA-binding proteins (see p. 26), but is otherwise unrelated to previously described proteins.

• *BRCA2* — mapped to 13q, accounts for a proportion of early onset breast cancer, and appears to equal that resulting from *BRCA1*, but *BRCA2* may not influence ovarian cancer risk.

A number of other genes that predispose to familial cancer, including the *p53* gene, probably account for the remaining susceptibility to early onset breast cancer.

The identification of susceptibility genes now allows genetic screening for predisposition to breast cancer. This raises a number of ethical issues, and careful counselling must precede any decision to screen for mutations. Testing reassures women who have not inherited a mutated gene, but the identification of women at risk has profound psychological implications for the patient, and raises difficult issues regarding the appropriate treatment.

Cancer therapy

Cancer vaccines

Tumours often provoke an immune response against themselves, as evidenced by the detection of circulating tumour-reactive T lymphocytes, and the presence of tumour-infiltrating lymphocytes. However, tumours can also protect themselves against immune reactions. For example, many tumours express only small amounts of MHC molecules which are required to present tumour-derived peptides to the hosts immune system.

A number of strategies are being developed to immunize patients against their cancers.

Immunization with tumour-specific peptides

Mutated oncogenes, such as *p53*, produce tumour-specific 'oncoproteins' that can generate a cytotoxic T-lymphocyte response. Peptide fragments of such proteins could therefore be used as vaccines to immunize patients against their cancers.

One particular group of tumours lends itself particularly well to this form of therapy. B-cell lymphomas arise through the clonal proliferation of a single B cell. Each B-cell lymphoma therefore expresses a distinct antibody, with a unique V region forming the antigen binding site, on its surface (see p. 114). Advances in recombinant DNA technology make it possible to clone the V-region gene from lymphoma cells, and design a recombinant vaccine that fuses the V-region protein with other proteins such as GM-CSF to make it more immunogenic.

Direct injection of a human leucocyte antigen (HLA) gene not expressed by the tumour

The gene for a foreign HLA antigen is injected into the tumour in a form in which it can be taken up by the tumour (e.g. as a DNA/liposomal complex, see p. 60). The hope is that in addition to recognizing the injected HLA antigen as foreign, an immune response will be generated against other tumour antigens.

Gene therapy for cancer

Several approaches to gene therapy for cancer are currently being pursued.

• Retroviruses can be used to deliver specific therapy to tumour cells. For example, an enzyme involved in the activation of a cytotoxic drug can be linked to the promoter region of a gene that is preferentially expressed in the tumour.

• Tumour-infiltrating lymphocytes (obtained by biopsy of the tumour) can be infected with retroviruses containing human cytokine genes, such as tumour necrosis factor or interleukin 2. The lymphocytes are then reintroduced into the patient in an attempt to deliver high concentrations of cytokines to the tumour. An alternative approach is to transfect the cytokine gene into tumour cells, which are then reintroduced in the hope that cytokine secretion by the tumour cells will activate an immune response.

• Tumour-suppressor genes or oncogenes could be manipulated within cancer cells. For example, genes such as *p53* could be introduced into cancer cells. Alternatively, retroviruses could be used to deliver antisense oligonucleotides (DNA sequences which bind to messenger RNA (mRNA) preventing its translation into protein, see p. 76), which interfere with the expression of mutant oncogenes.

Infectious diseases

Progress in molecular biology has relied on the use of viruses, yeast and bacteria. These tools are now extensively used in the diagnosis and eradication of these microorganisms.

Diagnosis of infections

PCR

The diagnosis of an infection usually requires culture and identification of the suspected infectious agent from samples obtained from the patient. Culture can be difficult or slow, resulting in inevitable delays in obtaining a diagnosis. This is particularly the case with certain bacteria (e.g. tubercle bacilli and rickettsiae) and most viruses. PCR can be used to amplify DNA (or viral RNA) sequences that are specific for an infectious agent, allowing the rapid diagnosis of infections.

It is a useful tool in the diagnosis of viral infections, including hepatitis B and C, human immunodeficiency virus (HIV) and HSV, and should prove of value in the diagnosis of bacterial infections such as tuberculous meningitis which can prove difficult to confirm using current bacteriological techniques.

Monoclonal antibodies

Monoclonal antibodies recognize target antigens with tremendous specificity and precision (see p. 115). This makes them ideal to detect specific external antigens on microorganisms. Examples include *Legionella*, cytomegalovirus and herpesviruses.

Monoclonal antibody therapy in septic shock. Septic shock is a systemic response to infection which carries a high mortality despite advances in antimicrobial therapy and intensive care practice. Various inflammatory mediators, including the cytokine tumour necrosis factor and extracts of bacterial cell walls (endotoxins), have been identified as mediators of septic shock. This has prompted clinical trials of monoclonal antibodies that neutralize these agents. Although animal studies showed beneficial effects of these agents, results of clinical trials in patients have been less encouraging. Antibodies against tumour necrosis factor did not confer any survival benefit when used in human subjects with septic shock. Anti-endotoxin antibodies have been shown to be of benefit in a subset of patients with septic shock, but it may not be possible to predict which patients will benefit prospectively from therapy, and these agents have yet to enter routine clinical practice.

Vaccines

Vaccination against infectious diseases has been one of the greatest achievements of medicine, leading to the eradication of smallpox, and the control of other viral infections (including poliomyelitis, mumps, measles and rubella) and bacterial infections (including whooping cough, tuberculosis, tetanus and diphtheria).

Vaccination involves the administration of:
- live infectious agents, which have been attenuated to render them harmless (e.g. the combined measles, mumps and rubella vaccines, vaccines against poliomyelitis and typhoid); or
- killed preparations derived from the plasma of infected individuals (e.g. the original hepatitis B vaccine) or culture of the microorganism (e.g. the hepatitis A vaccine).

These preparations have potential disadvantages. Attenuated agents may rarely cause disease, whereas preparing vaccines from infected plasma carries the risk of transmitting other infections. Furthermore, it may prove difficult to culture the microorganism in sufficient quantities to make an inactivated product. Alternatively, it may be impossible to provide the necessary attenuation whilst retaining the ability to replicate without causing disease.

Recombinant proteins as vaccines

Recombinant DNA technology provides the opportunity to clone and express genes from an infectious agent, and obtain recombinant protein products that can then be tested as vaccines. However, considerable difficulties have been encountered in expressing proteins in soluble forms which are immunogenic. This in part reflects the fact that the protein may be folded into a configuration which fails to induce the relevant immune response.

The only successful recombinant vaccine which has been produced to date is the *hepatitis B vaccine*. The surface protein of the hepatitis B virus, when expressed in yeast cells, forms particles which closely resemble the natural viral protein. These recombinant particles provoke a strong immune response against the hepatitis B virus.

Peptides as vaccines

The development of an immune response to an infection depends upon the agent containing molecules which are recognized as foreign. Substances which provoke immune responses are known as antigens. Antigens can contain several sites or determinants, known as epitopes, that are capable of stimulating a protective immune response.

The simplest, and potentially safest, form of vaccine is therefore a peptide that is representative of an epitope present on the surface of the microorganism. With the availability of DNA sequencing techniques (see p. 70), the nucleic acid sequence of genes encoding potentially immunogenic proteins can be defined, and the amino acid sequence of peptides can be deduced. However, very few epitopes consist of linear amino acid sequences. Usually, they are composed of discontinuous amino acid sequences that are brought together by folding of the protein. Identification of potentially immunogenic epitopes has therefore provided a major challenge. (All T cell epitopes are linear (peptides).)

Recombinant infectious vector vaccines

An alternative approach is to insert DNA from an infectious agent into an attenuated bacterial or viral vector. Following infection with the recombinant product, an immune response is generated against both the vector and the inserted gene product.

A variety of attenuated bacteria and viruses (including *Salmonella* and vaccinia, respectively) are currently under investigation as live vaccine vehicles.

HIV infection

The development of therapies for AIDS requires an understanding of how HIV-1 integrates into the human genome, and how viral replication and viral gene expression are regulated.

The proviral genome of HIV-1 is 9–10 kbp long, and has three main structural genes.
• The *gag* (group-specific antigen) gene encodes the core protein antigens of the virion (intact virus particle) (Fig. 5.6). These are formed as the cleaved products of a larger precursor protein.

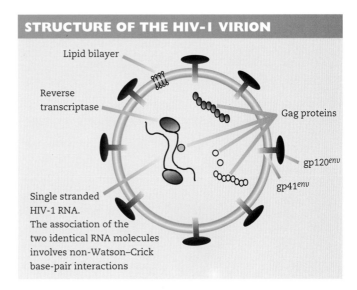

STRUCTURE OF THE HIV-1 VIRION

Lipid bilayer

Reverse transcriptase

Gag proteins

gp120^{env}

gp41^{env}

Single stranded HIV-1 RNA. The association of the two identical RNA molecules involves non-Watson–Crick base-pair interactions

Fig. 5.6 The structure of the HIV-1 virion.

• The *pol* (*poly*merase) gene encodes the viral reverse transcriptase, and also the IN protein required for *integration* of viral DNA into the host genome.
• The *env* gene encodes the two *env*elope glycoproteins, which are cleaved from a larger precursor.

In addition, a number of other genes encode protein products.

• The *pro* gene encodes the protease that cleaves the gag and pol protein precursors.
• The *vif* gene encodes a protein necessary for virion *inf*ectivity.
• The *nef* gene product functions as a *negative factor* for viral replication.
• The *tat* gene encodes a protein with *trans*-activating function. It interacts with the *trans*-activation response (TAR) region of the 5′-long terminal repeat (LTR), which contains regulatory sequences involved in viral gene expression.
• The *rev* gene encodes the *regulator of virion protein*, which determines whether structural genes are spliced from the viral mRNA, thereby determining whether only regulatory proteins or complete viral particles ready for nuclear export are made (Fig. 5.7).

When the HIV virion binds to a CD4 molecule on the cell surface a conformational change occurs in the envelope glycoprotein, and the virus enters the cell via fusion of lipid bilayers at the cell surface. The uncoated core of the virion then uses its viral reverse transcriptase to transcribe one of

Fig. 5.7 Diagrammatic representation of the HIV-1 genome. The 5′- and 3′-LTRs contain regulatory sequences. Several genes overlap.

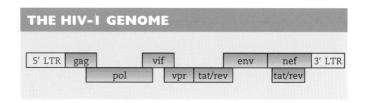

THE HIV-1 GENOME

the two identical strands of positive sense RNA into DNA. This DNA is duplicated by a host cell DNA polymerase, and migrates to the nucleus where it is integrated at a random site into the genome. Transcription of the integrated viral DNA is regulated by both host factors (such as the DNA binding protein NFϰB), and viral regulatory proteins such as the tat and rev proteins. Virally encoded proteins are processed and assembled in the cytoplasm, and then bud from the cell surface as new infectious virions.

Therapy for AIDS

Despite increased understanding of the molecular biology of HIV infection, therapies remain elusive and AIDS remains a progressive, fatal disease. Several sites in the viral life cycle have been targeted.

• *HIV entry*—soluble truncated forms of the CD4 molecule are effective at neutralizing HIV *in vitro*. However, clinical trials of *soluble recombinant CD4* failed to show any benefit in HIV-infected patients. Several factors may have accounted for this lack of efficacy. Recombinant CD4 has a short half life *in vivo*, and administration leads to the development of anti-CD4 antibodies. In addition, viruses present in the community may be less susceptible to neutralization by CD4 than their laboratory counterparts.

• *Inhibition of reverse transcriptase*—remains the mainstay of treatment. Zidovudine, and the newer agents didanosine (ddI) and dideoxycytosine (ddC) are nucleoside analogues which bind preferentially to viral reverse transcriptase compared to human DNA polymerase.

• *Prevention of new virion formation*—a number of treatments aimed at inhibiting the production of new virus particles are under evaluation. These include agents that bind to and inhibit viral proteins such as tat, and the *pro* gene product.

HIV vaccines

Emphasis is currently being placed on the development of vaccines against HIV. In addition to their importance in prevention, vaccines may also induce additional immunity in patients which is additional to any generated as a consequence of natural infection.

A number of therapeutic trials using either modified whole virus particles or recombinant HIV proteins are currently in progress. The majority of current vaccines are based on the extracellular envelope protein gp120, or the envelope precursor protein gp160. One factor which may limit the success of these vaccines is the high variability of the envelope proteins between different strains of HIV.

Gene therapy for AIDS

• The gene encoding an HIV protein, such as the gp120 envelope protein, is introduced into the patient, either directly into muscle cells, or *ex vivo* into cells that are harvested and reimplanted. Expression of the protein could then stimulate an immune response against the virus.

• Cells could be genetically engineered to secrete a potential therapy into the circulation. For example, lymphocytes could be transfected with a soluble CD4–IgG chimeric protein in the hope that continued secretion of soluble CD4 may be more effective at neutralizing HIV than injection of the recombinant molecule (see p. 140).

• Genes could be targeted directly to infected CD4 cells using retroviral vectors. For example, expression of a mutant *rev* gene could inhibit the production of structural HIV proteins. Alternatively, the expression of high levels of sequences that will bind HIV transcription factors (such as the TAR), could sequester the viral transcription factors.

Transplantation

HLA typing

HLAs are a family of glycoproteins that comprise the MHC in humans and are encoded on the short arm of chromosome 6. There are three HLA class I molecules, known as HLA-A, HLA-B and HLA-C, that are expressed on the surface of all nucleated cells, and three HLA class II molecules, known as HLA-DR, HLA-DP and HLA-DQ, that are normally expressed on a much smaller subset of cells, particularly lymphocytes, although their expression on other cell types can be induced during immune reactions. The function of both classes of HLA molecules is to bind pieces of proteins (peptides) and display them on the cell surface for recognition by T lymphocytes.

HLA class I molecules bind peptides derived from intracellular parasites, such as viruses, and present them to cytotoxic T cells, which kill virally infected cells. HLA class II molecules bind peptides derived from extracellular pathogens and present them to helper T cells, which help B lymphocytes to produce antibodies.

The genes encoding HLA antigens are extremely polymorphic, meaning that a large number of different alleles exist for each gene. Thus, the pattern of HLA molecules expressed, known as the 'HLA type', varies considerably between individuals. Whilst this is probably important for the health of the population because of the broader resistance to pathogens, it makes organ transplantation much more difficult because the recipient mounts a strong immune response against any different HLA molecules expressed on the donor tissue. Clinically the extent of match between HLA molecules present on the organ transplant and the recipient is an important determinant of the survival of the organ graft. Previous methods of HLA typing required days of culture or large panels of antisera, and involved complicated subsequent analysis. PCR provides a rapid method of HLA typing (known as tissue typing) prior to transplantation, by testing different HLA-specific primers to see if they yield an amplification product.

Monoclonal antibody therapy in transplant rejection

T cells are the principal mediators of transplant rejections. Monoclonal antibodies against T-cell subsets have been extensively used to identify T cells in rejecting organs, and improve understanding of the rejection process. In addition, many highly specific monoclonal antibodies against T-cell antigens have been developed as potential therapies for rejection. The most widely used has been the mouse monoclonal antibody OKT3, which recognizes CD3 (CD stands for cluster of differentiation), which forms part of the human T-cell receptor. Many patients develop an antimouse antibody response which limits the use of this agent, but antibody engineering to 'humanize' OKT3 has been undertaken.

Molecular biology is of fundamental importance in laboratory science, commerce and clinical practice. Although motivated by potentially diverse interests, the ultimate goal in these fields is the development of new therapies that will benefit patients. The impact of molecular biology is at last beginning to translate into such benefits.

Index